Manual of FIELD BIOLOGY and ECOLOGY

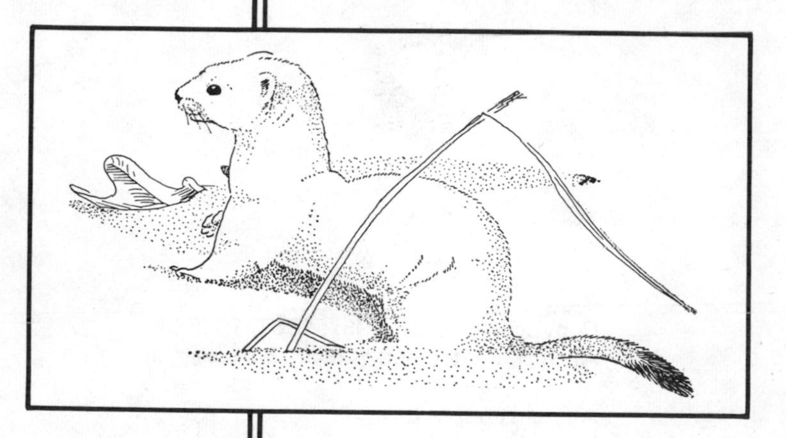

by

Allen H. Benton
Biology Department
State University College
Fredonia, New York

and

William E. Werner, Jr.
Biology Department
Blackburn College
Carlinville, Illinois

Burgess Publishing Company
426 South 6th Street • Minneapolis, Minn. 55415

Printed in the United States of America
Standard Book Number 8087-0217-3

Preface to the Fourth Edition

It was our belief, when we produced the first edition of the workbook, that there was a need for a manual which would provide guidance and direction for the beginning student, while including references and suggestions which would prove useful when the student had advanced in his biological studies. The reception it has had in colleges, outdoor laboratories and camps throughout the country has been most gratifying, and shows that it has filled a genuine need.

This fourth edition departs less from the form of the previous edition than have the earlier revisions. We have retained the sections of the third edition, adding new exercises in only a few cases. We have made major changes only in the literature section, where we have tried to bring up to date this important part of the book. Other sections have undergone only minor changes.

This manual is designed to accompany "Field biology and ecology", by the same authors, published by McGraw-Hill Book Co., New York. Reading to accompany the exercises will be helpful in broadening the student's understanding of the subject matter.

Among the many from whom we have gained ideas which have been used in writing exercises herein, we are especially grateful for permission to use exercises adapted from material provided by Dr. Lamont C. Cole, Cornell University, Dr. William J. Hamilton, Jr., Cornell University, and Dr. Arther S. Einarsen, Oregon State College.

We are particularly grateful to H. Wayne Trimm and the New York State Conservation Department, for permission to use the cover illustration and the many small line drawings which appear throughout the book. William Saulsberry, State University College at Fredonia, has been helpful in various artistic details.

Introduction

Modern biology has come to mean, primarily, molecular and cellular biology. Tremendous advances in instrumentation and technique have revolutionized these areas, and have made them the forefront of advance in biology. Nonetheless, the broad spectrum of knowledge which may be included under the heading of field biology has not come to a standstill. New instrumentation and new techniques have been as important here as in laboratory biology. Further, the number of people with an amateur interest in field biology has continued to increase. Bird clubs, conservation groups, natural history societies and other related organizations have grown and flourished. Meanwhile, the number of people professionally engaged in field work of a biological nature has also increased, with expanding conservation departments, entomological bureaus, and other governmental agencies providing the biggest impetus.

Most of these people, professional or amateur, use similar techniques and require a similar background of knowledge. Each must be able to recognize the organisms with which he works, and must know something of their distribution, their ecological requirements, their way of life. Each must know how to collect and preserve samples of these organisms, where and when to find them, and from whom to get additional information about them. Each must have a sound knowledge of the voluminous literature of his particular field, and must know how to use varied reference books. Each must know how to observe, and how to take and interpret notes on his observations. Each must know how to use such tools as cameras, binoculars, recording devices, chemical materials etc. as the exigencies of his particular profession or hobby require.

It is the function of this manual to provide a broad basis for obtaining the kinds of knowledge indicated above. It is not intended as a definitive compendium nor as a sole textbook. With this book as a companion and guide, the beginner may hope to enter the distinguished company of such field biologists as Linnaeus, Darwin, Fabre and Audubon, and to find pleasure and profit in the observation, study and contemplation of nature.

A note to the instructor: Since this manual is intended to be suitable for use anywhere in the United States and Canada, and since field trips must vary according to the time of year and the climate, as well as the types of environments available, many more exercises are included than can be used in one course. Similarly, the indoor exercises are so arranged as to allow for either expansion or contraction, at the discretion of the teacher. We recognize that the literature lists are more extensive than many beginning students will require, but we feel that the student will find them more valuable as his interest and skill develop. Furthermore, their inclusion permits use of the book at varied levels of competence.

Some instructors have found it desirable to divide larger groups of students into teams. Thus the duties of collecting data, taking notes from the instructor, gathering and transporting specimens may be divided among team members. Duplicate data sheets may be prepared by the members of a team, so that one may be torn out and handed in, the others retained for student use.

This book has been designed for convenience and utility in field and laboratory. It is our hope that it will provide as many pleasant hours for its users as it has provided for its authors during its preparation.

TABLE OF CONTENTS

* An asterisk before an entry indicates a possible laboratory exercise or field trip.

Page

SECTION VII - PROJECTS FOR FIELD STUDY

SECTION VIII - SELECTED BIOLOGICAL LITERATURE

Section I
Taxonomy

Every biologist, regardless of his area of interest, must learn the fundamentals of classification. Indeed, the first step in any study is the correct identification of the organisms being studied, and this is a matter of taxonomic procedure.

Our system of classification, known as the binominal (two name) system, was originated by a Swedish biologist named Karl von Linné in the mid-eighteenth century. Each organism was given a scientific name consisting of two Latin (or Latinized) words, followed by the Latinized name of the describer. Hence von Linné is best known to us today by his Latinized name, Linnaeus. The first word of the scientific name, always written with a capital, is the generic name, that is, the name of the genus in which the organism is placed. The second word is the specific or trivial name, which is ordinarily not capitalized. Thus the mallard duck, which occurs almost throughout the world, may be known by many common names in various countries, but its scientific name, Anas platyrhynchos, is the same everywhere. In many cases, a third word has been added as subspecies have been described, and in these cases the name is trinominal.

Each species belongs to a series of higher categories, also originated by Linnaeus. This hierarchical classification includes, from greatest to smallest, the following categories:

Kingdom
 Phylum
  Class
   Order
    Family
     Genus
      Species

Various sub- and super-groups (e.g., superfamily, subclass) have been added, but these are ordinarily used only by the specialist. For the beginner, a knowledge of the characteristics of the various groups, and a degree of facility in their identification, is all that is to be expected. The references listed below, the keys listed on p. 11 and the identification guides listed on pp. 218-226 will aid in this accomplishment.

REFERENCES

Copeland, Herbert F. 1956. Classification of lower organisms. Pacific Books, Publ., Palo Alto, Calif.

Follett, W. I. 1963. New precepts of Zoological Nomenclature. A.I.B.S. Bull. 13:14-18.

Mayr, Ernst, E. G. Linsley, and R. L. Usinger. 1953. Methods and principles of systematic zoology. McGraw-Hill Book Co. Inc., New York.

Schenk, Edward T., and John McMasters. 1956. Procedure in Taxonomy. 3rd Ed. Stanford Univ. Press, Calif.

Simpson, George G. 1961. Principles of animal taxonomy. Columbia Univ. Press, New York.

Classification of Existing Organisms

The human race being what it is, we have not arrived at any one generally accepted scheme of classification. For many years, it has been customary to divide organisms into two kingdoms, Plantae and Animalia, each divided into phyla. In recent years, there has been some opinion in favor of adding other groups, at least roughly equivalent to kingdoms, primarily because of the fact that many one-celled (or acellular) organisms are not readily placed into either the plant or animal kingdom. One such scheme places all one-celled animals in a kingdom, Protista, emphasizing that primitive organisms were neither plant nor animal, and avoiding the necessity of placing current forms in either one. Another scheme adds still another kingdom, the Monera, to include the bacteria and the blue-green algae, which are considered by many authorities to be quite closely related and quite distinct from all other existing groups.

We have chosen to use a relatively conservative classification, pointing out where necessary certain areas of disagreement. In spite of the disquieting aspects of a classification which is not entirely stable, and not universally accepted, we must accept the fact that progress in the understanding of relationships must result in progress in the arrangement of groups, unless we abandon completely the idea of a classification which will in some measure reflect the real relationships of organisms.

In addition to the references listed above, there have been

in recent years a number of books introducing new or modified techniques into the practice of taxonomy. Taxonomic decision based on morphological characteristics alone is still the main method of operation, but other characteristics and different methods are detailed in the two books here listed.

REFERENCES

Sokal, Robert P., and Peter H. A. Sneath. 1963. Principles of numerical taxonomy. W. H. Freeman Co., San Francisco.
Swain T., Ed. 1963. Chemical plant taxonomy. Academic Press, New York.

A Classification of the Animal Kingdom

PHYLUM PROTOZOA - unicellular animals
- Class Sarcodina - Amoeba and relatives
- Class Mastigophora - Flagellates
- Class Sporozoa - Malarial organisms and other parasites (sporozoans)
- Class Ciliata - Ciliates
- Class Suctoria - Suctorians

PHYLUM PORIFERA - Sponges
- Class Calcarea - Limy sponges
- Class Hexactinellida - Glass sponges
- Class Demospongiae - Bath sponges and relatives

PHYLUM COELENTERATA - Coelenterates
- Class Hydrozoa - Hydra and relatives
- Class Scyphozoa - Jellyfishes
- Class Anthozoa - Corals, sea anemones

PHYLUM CTENOPHORA - Sea walnuts, comb jellies
- Class Tentaculata
- Class Nuda

PHYLUM PLATYHELMINTHES - Flatworms
- Class Turbellaria - Free-living flatworms
- Class Trematoda - Flukes
- Class Cestoda - Tapeworms

PHYLUM NEMERTINEA - Ribbon worms

PHYLUM ASCHELMINTHES - Roundworms and relatives
- Class Rotatoria - Rotifers*
- Class Gastrotricha - Gastrotrichs
- Class Echinodera -
- Class Nematoda - Roundworms
- Class Nematomorpha - Horsehair worms

PHYLUM ACANTHOCEPHALA - Spiny headed worms

PHYLUM ENTOPROCTA -

PHYLUM BRYOZOA - Moss animals

PHYLUM PHORONIDA - Phoronids

PHYLUM BRACHIOPODA - Lamp shells

PHYLUM SIPUNCULOIDEA - Peanut worms

PHYLUM PRIAPULOIDEA -

PHYLUM CHAETOGNATHA - Arrow worms

PHYLUM ECHIUROIDEA -

PHYLUM ANNELIDA - Segmented worms
- Class Archiannelida - Primitive marine worms
- Class Polychaeta - Clamworms and relatives
- Class Oligochaeta - Earthworms and relatives
- Class Hirudinea - Leeches

PHYLUM ARTHROPODA - Arthropods
- Class Crustacea - Crayfish and relatives
- Class Onychophora - Peripatus*
- Class Chilopoda - Centipedes
- Class Diplopoda - Millipedes
- Class Arachnida - Spiders, ticks and relatives
- Class Insecta - Insects**

PHYLUM MOLLUSCA - Molluscs
- Class Amphineura - Chitons
- Class Gastropoda - Snails and slugs
- Class Scaphopoda - Tooth shells
- Class Pelecypoda - Clams and relatives
- Class Cephalopoda - Squids and relatives

* Placed by some authorities in a separate phylum.
** For key to orders of insects, see page 12.

PHYLUM ECHINODERMATA - Spiny-skinned animals
- Class Crinoidea - Sea lilies
- Class Asteroidea - Starfishes
- Class Ophiuroidea - Brittle stars
- Class Echinoidea - Sea urchins
- Class Holothuroidea - Sea cucumbers

PHYLUM CHORDATA - Chordates
- Subphylum Hemichordata - Acorn worms
- Subphylum Urochordata - Sea squirts
- Subphylum Cephalochordata - Lancelets
- Subphylum Vertebrata - Vertebrates
  - Class Agnatha - Lampreys and relatives
  - Class Chondrichthyes - Sharks and relatives
  - Class Osteichthyes - Bony fishes
  - Class Amphibia - Frogs, toads, salamanders and caecilians
  - Class Reptilia - Turtles, crocodilians, snakes and lizards
  - Class Aves - Birds
  - Class Mammalia - Mammals

REFERENCES

Blackwelder, Richard E. 1963. Classification of the animal kingdom. Southern Illinois Univ. Press, Carbondale, Ill.
Blair, W. H. 1961. Vertebrate speciation. Univ. of Texas Press, Austin, Tex.
Borradaile, L. S., and F. A. Potts. 1958. The Invertebrata. 3rd Ed. Cambridge Univ. Press, Cambridge, Eng.
Calman, W. T. 1949. The classification of animals. Methuen Co., London, Eng.
Hyman, Libbie H. 1940 et seq. The invertebrates. Vols. 1-5. McGraw-Hill Book Co., N. Y.
Mayr, Ernst. 1963. Animal species and evolution. Belknap Press of Harvard Univ. Press, Cambridge, Mass.
Rothschild, Nathanael M. V. 1961. Classification of living animals. John Wiley & Sons, N. Y.

A Classification of the Plant Kingdom

PHYLUM CYANOPHYTA - Blue-green algae

PHYLUM CHLOROPHYTA - Green algae

PHYLUM CHRYSOPHYTA - Diatoms

PHYLUM PHAEOPHYTA - Brown algae

PHYLUM RHODOPHYTA - Red algae*

PHYLUM SCHIZOMYCOPHYTA - Bacteria

PHYLUM MYXOMYCOPHYTA - Slime molds

PHYLUM EUMYCOPHYTA - Fungi
- Class Phycomycetae - Molds and relatives
- Class Ascomycetae - Cup fungi, yeasts and relatives
- Class Basidiomycetae - Club fungi, mushrooms, puffballs and relatives
- Class Fungi Imperfecti - Fungi not classifiable according to present knowledge

PHYLUM BRYOPHYTA - Mosses and liverworts
- Class Hepaticae - Liverworts
- Class Musci - Mosses
- Class Anthocerotae - Hornworts

PHYLUM PSILOPSIDA - Naked ferns**

PHYLUM LYCOPSIDA - Clubmosses and relatives

PHYLUM SPHENOPSIDA - Horsetails and relatives

PHYLUM PTEROPSIDA - Vascular plants
- Class Filicinae - Ferns
- Class Gymnospermae - Conifers and relatives
- Class Angiospermae - Flowering plants
  - Subclass Monocotyledonae - Monocots
  - Subclass Dicotyledonae - Dicots

REFERENCES

Benson, Lyman. 1964. Plant classification. D. C. Heath Co., Boston.

Bessey, Ernst A. 1950. Morphology and taxonomy of fungi. McGraw-Hill Book Co., N. Y.

Core, Earl L. 1955. Plant taxonomy. Prentice-Hall, Englewood Cliffs N. J.

Kessel, Edward L. (Ed.) 1955. A century of progress in the natural sciences. 1853-1953. Calif. Acad. Sci. San Francisco, Calif.

* Three additional phyla are sometimes inserted among the algae: Pyrrhophyta, the dinoflagellates and their relatives; Euglenophyta, the euglenoid flagellates; and Charophyta, the stoneworts.

** Some authorities place this and the following phyla in the single phylum Tracheophyta, reducing the groups here listed as phyla to sub-phyla.

Lawrence, George H. M. 1951. Taxonomy of vascular plants. Macmillan Co., N. Y.
Smith, Gilbert M. 1950. The freshwater algae of the United States. McGraw-Hill Book Co., N. Y.
_______________ 1955. Cryptogamic botany. Vols. 1 & 2. McGraw-Hill Book Co., N. Y.
Sokal, Robert R., and Peter H. A. Sneath. 1963. Principles of numerical taxonomy. W. H. Freeman Co., San Francisco.
Swain, T., Ed. 1963. Chemical plant taxonomy. Academic Press, N. Y.

The Origin and Meaning of Scientific Names

The beginning biologist is often confused by the complicated names which have been given to the organisms he must study. Though it is easy to point out that he already knows many such names (Geranium, Gardenia, Trillium, Begonia, Hippopotamus, Rhinoceros), the names still look forbidding. It is often helpful to see where the name came from, but since modern education often includes no Latin or Greek the student is ill-prepared to recognize the component parts of scientific names.

As a brief introduction to this interesting side avenue of biology, the following list has been compiled. It contains the most commonly used descriptive terms, and a variety of combining terms which are found in many scientific names. Though this list is obviously not exhaustive, the interested student may find a more complete list in one of the books mentioned below.

Generic names are often derived from the classical languages. Thus the Greek word for squirrel, Sciurus, remains the generic name of some of our squirrels, while the Latin Equus is still applied to our horse. Other common derivations are from names (Linnaea, Franklinia, etc.), from obvious characteristics (Hippopotamus means river-horse, Dolichopus means long leg), and various forms of "made-up" names such as Blarina and Dafila. The trivial name is most often descriptive, but may also be geographical, (americana, californiense, etc.), dedicated to someone (auduboni, holbrooki, bairdi, etc.), or may be derived in some other way. In all cases the name is Latinized and the rules of Latin grammar apply to its use.

DESCRIPTIVE TERM	ENGLISH MEANING
aestivalis	of summer
albus	white
agrarius	of the field

DESCRIPTIVE TERM	ENGLISH MEANING
aquatilis	of water
argenteus	silvery
arvensis (arvalis)	of the field
atratus	blackened
borealis	northern
caeruleus	blue
carinatus	keeled
caudatus	tailed
coccineus	scarlet
concolor	all of one color
cristatus	crested
cyaneus	blue
discolor	of more than one color
echinatus	prickly or spiny
ferox	fierce
flabellatus	fan-like
fluviatilis	of the river
furcatus	forked
gracilis	slender
humilis	dwarfed, low-growing
hyemalis	of the winter
lacustris	of the lake
litoralis	of the seashore
maculatus (maculosus)	spotted
maritimus	of the sea
montanus	of the mountains
nigricans	black
nitidus	shining
nivalis	snowy or of the snow
noveboracensis	of New York
occidentalis	western
orientalis	eastern
palustris	of the marsh
parvulus	very small
pectinatus	comb-like or combed
pratensis	of meadows
pulchellus	beautiful
pusillus	insignificant
riparius	of river banks
rostratus	beaked
saxicolus	inhabiting rocks
septentrionalis	northern
stellatus	starry
tenebrosus	of dark places
tenuis	thin
unguiculatus	clawed

DESCRIPTIVE TERM	ENGLISH MEANING
venosus	veiny
vernalis	of spring
virens	green
vulgaris	common

NUMBERS

mono-	one
bi-, di-	two
tri-	three
quadra-	four
quinque-	five
sex-	six
septem-	seven
octo-	eight
novem-	nine
deca-	ten

Numbers are normally used in combination, e.g., quadramaculatus (four spotted), triunguis (three nails or claws), novemcinctus (nine banded), sexguttata (six spotted).

WORDS OR PREFIXES NORMALLY USED IN COMBINATION

an- or a-	without
auris (Lat.), otis (Gr.) . . .	ear
bios	life
brevis (Lat.) brachys (Gr.) .	short
cauda (Lat.) urus (Gr.) . . .	tail
caput (Lat.) cephale (Gr.) . .	head
chloros	green
chrysos	golden
dactylos	finger or toe
dermis	skin
dolichos (Gr.) longus (Lat.) .	long
erythros	red
flavus	yellow
hemi-	half
hydro	water
hyper-	above
hypo-	below
micro- (Gr.) parvus (Lat.) .	small
macro- (Gr.) grandis (Lat.) .	large
pachys	thick
philos	loving
platys	flat

WORDS OR PREFIXES NORMALLY USED IN COMBINATION

poly- (Gr.) multi- (Lat.). . .	many
pes (or pus or podus)	foot
pseudo-	false
rhinos	nose
stoma	mouth
trichos	hair

REFERENCES

Borror, Donald J. 1960. A dictionary of word roots and combining forms. N-P Publ., Palo Alto, Calif.

Brown, Roland W. 1954. Composition of scientific words. Published by the author, Baltimore, Md.

Henderson, I. F., and W. D. Henderson. 1963. Dictionary of biological terms. 8th Ed., edited by John Kenneth. D. Van Nostrand Co., Princeton, N. J.

Jaeger, Edmund C. 1950. Source-book of biological names and terms. C. C. Thomas, Springfield, Ill.

Jaeger, Edmund C. 1960. The biologist's handbook of pronounciation. C. C. Thomas, Springfield, Ill.

Jordan, David S. 1929. Manual of the vertebrate animals of the northeastern United States. World Book Co., N. Y.

Nybakken, O. E. 1959. Greek and Latin in scientific terminology. Iowa State Univ. Press, Ames, Iowa.

Savory, Theodore. 1963. Naming the living world. John Wiley and Sons, Inc., New York.

How to Use a Key

You will be given a number of organisms with which to work, and a key to the particular group of organisms which they represent. With one organism at hand, read the first couplet of the key. One of the statements in this couplet will agree with your specimen, and at the end of that statement you will find either (1) the name of your organism, or (2) the next number to which you must go in the key. At this next number you will find another couplet, where you can again select the statement which agrees with your specimen. Continue in this manner until you have reached a statement which ends with the name of an organism.

Since to err is human, and keys are imperfect, check your identification with a reference or with the instructor. If you are in error, go back in the key to the couplet at which you think you might have been wrong, and try the other statement. After a reasonable amount of practice, you should be able to run all but the most difficult specimens through the key without undue difficulty.

For the benefit of instructors who may wish to use a ready-made key, we have included a key to the orders of insects. Many instructors may wish, however, to utilize some other group of organisms in this exercise. A few keys which might be of value in this connection are listed below.

REFERENCES

Benton, Allen H., and Margaret M. Stewart. 1964. Keys to the vertebrates of the northeastern states (excluding birds). Burgess Publ. Co., Minneapolis, Minn.

Blair, W. F., et al. 1965. Vertebrates of the United States, 2nd Ed. McGraw-Hill Book Co., N. Y.

Eddy, Samuel, and A. C. Hodson. 1955. Taxonomic keys to the animals of the north central states, exclusive of the parasitic worms, insects and birds. Burgess Publ. Co., Minneapolis, Minn.

Jaques, H. E., Ed. About 25 books in the Pictured-Key Series cover most natural groups and are published by William C. Brown Co., Dubuque, Iowa.

Stains, Howard J. 1962. Game biology and game management: a laboratory manual. Burgess Publ. Co., Minneapolis, Minn. Pp. 22-29, 37-54.

Key to the Larger Orders of Insects

This key to the insect orders is intended to serve the needs of those classes which require a key in the workbook. The order names used here follow "An Introduction to the Study of Insects", by Borror and DeLong.

For the sake of increased simplicity, we have omitted certain small orders of insects which are unlikely to be encountered by the beginning student, or which require specialized preparation for identification and thus are not likely to be used for an exercise of this type. The omitted orders are listed at the end of the key.

In a few orders, certain aberrant groups will not key readily to the proper place. In these cases, we have adopted one of two courses. If the group in question is rare, we have omitted it, and the order is marked with an asterisk. Otherwise, we have so arranged the key that that portion of the order will key out at another point in the key.

Beginning students invariably have difficulty in the determination of certain rather obscure characteristics. Where our experience indicates that this is true, we have so arranged the key that both branches will lead to the proper answer. For example, the mouthparts of the Mecoptera are of the chewing type, but they are at the end of a beak-like structure. Thus, the Mecoptera will key out correctly whether the student chooses, in Couplet II, to say that it has chewing mouthparts or a beak. It is hoped that these simplifications will help the student to avoid some of the frustrating experiences which are the usual lot of the beginner in the use of keys.

Check your final determination with an insect identification book (See pps. 221-222).

1. Winged insect - 2
1. Wingless insects - 22

2. Mouthparts specialized for chewing, no beak, tube, etc., - 3
2. Mouthparts not specialized for chewing - 16

3. Front wings hard and horny, may be much reduced - 4
3. Front wings not hard and horny - 6

4. Abdomen with forceps-like structure at tip - Dermaptera, Earwigs
4. Abdomen without forceps-like structure - 5

5. Front wings without apparent veins, may be furrowed or striated - Coleoptera, Beetles*
5. Front wings with apparent veins - Orthoptera, Grasshoppers, etc.

6. Front wings transparent, glass-like - 7
6. Front wings opaque, pigmented, leathery or hairy - 14

7. Wings with numerous crossveins and complex venation - 8
7. Wings with relatively few crossveins - 12

8. Abdomen with 2 or 3 long hair-like tails - Ephemeroptera, Mayflies*
8. Abdomen without hairlike tails - 9

9. Mouthparts at end of downturned beak - Mecoptera, Scorpionflies
9. Mouthparts not at end of a beak - 10

10. Antennae very short, not obvious - Odonata, Dragonflies and Damselflies
10. Antennae obvious - 11

11. Tarsi three-segmented - Plecoptera, Stoneflies
11. Tarsi five-segmented - Neuroptera, Nerve-winged insects

12. Junction of thorax and abdomen constricted - Hymenoptera, Bees, etc.
12. Junction of thorax and abdomen not constricted - 13

13. Wings extending well beyond abdomen when at rest - Isoptera, Termites
13. Wings not extending far beyond abdomen - Hymenoptera

14. Wings covered with hair - Trichoptera, Caddisflies
14. Wings not covered with hair - 15

15. Junction of thorax and abdomen constricted - Hymenoptera
15. Junction of thorax and abdomen not constricted - Orthoptera

16. Only two wings present - Diptera, Flies*
16. Four wings present - 17

17. Mouthparts a coiled tube - Lepidoptera, Moths and Butterflies*
17. Mouthparts a beak - 18

18. Beak stout with chewing mouthparts at end - 19
18. Beak sharp and slender, adapted for piercing and sucking - 20

19. Wings membranous - Mecoptera
19. Forewings hard and horny - Coleoptera (weevils and relatives)

20. Wings fringed with hair - Thysanoptera, Thrips
20. Wings not fringed with hair - 21

21. Beak arising from front part of head - Hemiptera, True Bugs*
21. Beak arising from rear part of head - Homoptera, Leafhoppers, etc.

22. Peg-like structure on ventral side of abdomen - Collembola, Springtails
22. No peg-like structure on abdomen - 23

23. Mouthparts specialized for chewing - 24
23. Mouthparts not specialized for chewing - 28

24. Hind legs enlarged, adapted for jumping - Orthoptera
24. Hind legs normal in size - 25

25. Forceps-like structure at tip of abdomen - Dermaptera
25. No forceps-like strucutre on abdomen - 26

26. Junction of thorax and abdomen constricted - Hymenoptera
26. Junction of thorax and abdomen not constricted - 27

27. Tarsi four-segmented - Isoptera
27. Tarsi five-segmented - Orthoptera

28. Mouthparts a coiled tube - Lepidoptera
28. Mouthparts not a coiled tube - 29

29. Mouthparts not beak-like - Diptera
29. Mouthparts beak-like - 30

30. Beak stout, with chewing mouthparts at end - Mecoptera
30. Beak slender, adapted for piercing and sucking - 31

31. Tarsi with one or two segments - Thysanoptera
31. Tarsi with three segments - 32

32. Beak arising from front of head - Hemiptera
32. Beak arising from rear part of head - Homoptera

The following relatively small orders are omitted from this key: Anoplura, Embioptera, Mallophaga, Protura, Psocoptera, Siphonaptera, Strepsiptera, Thysanura, Zoraptera.

Securing Identification of Unknown Organisms

No matter how skilled one may become, he will never become able to identify all kinds of organisms. In fields of special difficulty, or in fields outside the biologist's field of competence, it is often necessary to seek the help of an expert. There are four major kinds of organizations where such help may be found.

1. State Experiment Stations or Agricultural Colleges: plants of economic importance, weeds, insect pests, indeed any organism of agricultural importance, may be identified by these experts.
2. State Museums, Natural History Surveys, and similar groups: Although such organizations cannot employ experts in all fields, their staff will include competent botanists, zoologists and entomologists, and they are thoroughly familiar with most local organisms.
3. College biology departments: Most professors of biology are especially interested in some group of organisms and are willing to identify them for anyone. Such specialists often work with the above listed organizations and are consulted when their services are needed.
4. Federal Government scientists: The U. S. Department of Agriculture maintains a Division of Insect Identification, while biologists at the U. S. National Museum include authorities in many fields. An inquiry to one of these groups may be a last resort, and will usually produce the desired results.

With the possible exception of some college professors, all

STUDENT: DATE:

Specimen number	Identification

of the scientists named above consider identification of organisms as a part of their work. In general, it is good etiquette to send a letter first, asking if they can and will examine your specimens. It is also good etiquette to offer them any of the specimens they may wish to keep for their collections, or at least any duplicates.

Speciation in Vertebrates

To the beginning biologist, the concept of the species is not easy to grasp. Since world-renowned authorities still wrangle over the subject, this is not strange, but the basic concepts involved are really quite simple. We will try to state and explain them here, and to give you an exercise which will illustrate their use.

No two organisms are exactly identical. Differences result from their genetic makeup and from environmental effects. Except for identical siblings, vegetatively reproducing plants and other asexual forms, no two organisms have the same genetic makeup. These differences in genetic makeup are the basis of our classification system, but the degree of such differences is not easy to determine. When two organisms differ in enough genetic characters, they will not be able to produce viable offspring, or perhaps not any offspring at all. They are then said to be reproductively isolated, and are considered as belonging to different species.

In practice, however, it is not easy for the taxonomist to find out either the degree of genetic difference or the degree of reproductive isolation. For this reason, our taxonomists use visible structural differences (morphological characteristics) in separating organisms, and decide whether organisms represent different species on the basis of these differences. Since morphological differences reflect genetic differences, this method is reasonably successful, although cases are known in which genetic differences can be seen by cytological study even though no consistent external structural differences have been found. Likewise, profound external differences may be the result of very minor genetic differences, often a single gene, and such differences may not affect interbreeding.

Modern taxonomists attack some of these difficult problems by a mathematical approach. Characteristics which will give numerical data (meristic characters) are used: the scales of fishes, measurements of tail, body length, etc. in mammals, wing length, tail length and similar measurements in birds. When large numbers of organisms are measured in this way, the measurements can be analyzed by statistical methods, and means may be found to show whether the populations involved represent different species. Let us consider some examples.

Over most of our country, the ring-necked snakes (*Diadophis* spp.) are common. In external appearance, they are very similar, and the southwestern form cannot be easily separated from the western form by superficial examination. The statistical approach, however, has shown that the southwestern population has more ventral scales than the western population, and they have thus been assigned as different species, the southwestern ring-necked snake, *Diadophis regalis*, and the western ring-necked snake, *Diadophis amabilis*.

When large samples of two populations show meristic characters in which there is no overlap, taxonomists usually consider them as distinct species. If there is overlap in all meristic characters, the populations may be subspecies of the same species, or merely populations with small and insignificant differences. Experience and taxonomic judgment are necessary for accurate evaluation of such cases.

Even the beginner, however, can perform the necessary mathematical calculations and make his own conclusion as to the taxonomic status of two populations. Perhaps the most widely used method is the one adopted by Hubbs and Perlmutter (1942) for taxonomic study of fish populations. This method involves the use of some relatively simple mathematical calculations, but it involves the use of terms which may not be familiar to everyone. Before doing this exercise, therefore, the following definitions will be necessary:

Mean: the arithmetic average of a series of measurements

Frequency: the number of times which a particular measurement occurs in a sample (represented by f)

Range: the difference between the extremes of measurement in a sample

Frequency distribution: a chart showing the frequency of each measurement within the range

Standard deviation: a mathematical measure of variability, expressed in terms of the units of measurements (represented by sd)

Standard error of the mean: a mathematical measure of the variability which might be expected in the means of different samples taken from a given population (represented by se)

$$sd = \sqrt{\frac{\Sigma fx^2 - (sfx)^2/N}{N - 1}}$$

$$se = \frac{sd}{\sqrt{N}}\ .$$

Number in sample (N) : 150

Range of sample: 46-56

Measure- ments	Fre- quency (f)	Deviation from mean (x)	fx	$f(x^2)$
46	2	-5	-10	50
47	3	-4	-12	48
48	10	-3	-30	90
49	15	-2	-30	60
50	12	-1	-42	42
51	25	0	0	0
52	23	1	23	23
53	14	2	28	56
54	11	3	33	99
55	3	4	12	48
56	2	5	10	50
			sfx: -18	Σfx^2: 566

(sfx^2): 324

Substituting in the above formula for finding standard deviation:

$$sd = \sqrt{\frac{566 - 324/150}{149}} = 1.94$$

$$se = \frac{1.94}{\sqrt{150}} = .158$$

In comparing two or more populations, these data can be placed on graph paper as below. This is explained more fully on page 20.

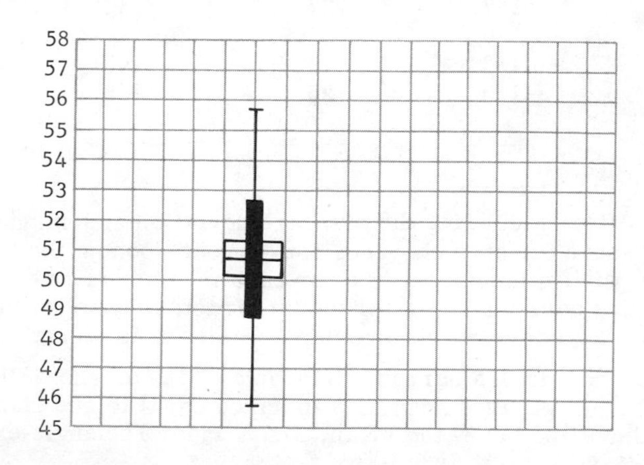

This is an actual problem, based on collections made in an area where an ancient eroded plateau rises from narrow valleys on both sides. Red-backed mice, Clethrionomys gapperi, occur both in the valley areas and on top of the plateau, but it appeared to a collector that those from the higher parts of the plateau were larger, brighter in coloration and heavier than those from other nearby areas. He therefore selected 40 adults, including no pregnant females, 20 from the top of the plateau and 20 from other collecting sites. Below are given the two sets of measurements. (Color, of course, is non-meristic without specialized instruments, and thus cannot be used here). Set up a frequency distribution, and figure standard deviation and standard error.

	Population A (non-plateau areas)		Population B (plateau)	
	Total Length	Weight	Total Length	Weight
	139 mm.	23 gr.	162	38.4
	136	23	155	36.5
	135	23.3	148	28.7
	133	23.8	153	33.6
	139	23.5	145	29.4
	130	21.6	141	25.8
	131	25	142	24.1
	131	22	149	32
	126	21.9	142	27.4
	135	23	159	28.8
	130	18.9	141	24.6
	126	20	142	26
	130	19.2	139	21.3
	135	20.5	149	30.4
	141	21.4	153	28.1
	138	25.3	138	22.5
	137	27	132	21.1
	140	23.9	147	24.4
	132	18.7	149	32.5
	125	25.1	145	36
Mean:	______	______	______	______
Range:	______	______	______	______

After calculating the standard deviation and standard error of the mean for each characteristic of each population, plot on graph paper a horizontal line for each mean, and a vertical line extending over the range for each characteristic of each population. Do the ranges for weight overlap? Do those for length overlap?

Next, draw a narrow rectangular box, as shown in the sample diagram on p. 19 extending a distance equal to one standard deviation above and below the mean, and a wider rectangle extending a distance equal to twice the standard error of the mean on each side of the mean.

The degree of overlap in these rectangles is a measure of the probability that the two samples came from the same population. Precise figuring of probability requires use of a t table which can be found in a standard set of statistical tables. In general, if the two rectangles show no overlap, it is highly improbable that the two samples came from the same population. If they show broad overlap, it is probable that they came from the same population. Do the wider rectangles overlap in either case? Do you think that the two samples actually came from the same population and that the apparent differences are of no taxonomic significance, or does it appear that the two populations are indeed significantly different?

A simpler device may be used to check your conclusions. This is known as the Rank Method, and is a rapid approximation method, which shows with a fair degree of accuracy the probability that two samples derive from different populations.

In this method the measurements are given a rank starting with the smallest. Thus, in the example above, the length of 125 would be given a rank of 1, and on up to 162, which would be given a rank of 40. If the same measurement occurs more than once, an average figure is used; thus 126 occurs twice, and these two measurements would be 2nd and 3rd in rank. They are thus ranked at 2.5 each, and the next figure, 130, which also occurs twice would be ranked as 4th and 5th and assigned 4.5 each.

The ranks for each sample are then totalled. If they are drawn from the same population, the two totals would be expected to be nearly the same. If the difference is marked, its significance may be calculated by the use of the table below. Find the smaller rank total in one of the two columns opposite the number of individuals in each sample (in this case, 20). If the number is greater than that in the P = 0.05 column, the results are not significant at the 95% level, the level usually required by such studies as this. If, on the other hand, the figure is smaller than that in the P = 0.05 column, the probability is that the two samples came from different populations.

Using the figures above, calculate the probability that these samples were drawn from the same population. How closely does this agree with results from the other method?

N	Significant at the 95% level P = 0.05	Significant at the 99% level P = 0.01
5	18	15
6	27	23
7	27	32
8	49	44
9	63	56
10	79	71
11	97	87
12	116	105
13	137	125
14	160	147
15	185	170
16	212	196
17	241	223
18	271	252
19	303	282
20	338	315

This table is reproduced from "Some rapid approximate statistical procedures", by Frank Wilcoxon, American Cyanamid Co., N.Y., 1949. Reproduced by permission of American Cyanamid Company.

REFERENCES

Amadon, Dean. 1943. Bird weights as an aid in taxonomy. Wilson Bull. 55:164-177.

Anderson, Edgar. 1949. Introgressive hybridization. John Wiley and Sons Inc., N. Y.

Barbehenn, Kile R., and John New. 1957. Possible natural intergradation between prairie and forest deer mice. J. of Mammal. 38:210-218.

Hubbs, Carl L., and Alfred Perlmutter. 1942. Biometric comparison of several samples, with particular reference to racial investigations. Am. Nat. 76:582-592.

Pyburn, William F., and J. P. Kennedy. 1960. Artificial hybridization of the gray treefrog, *Hyla versicolor*. Am. Midland Naturalist. 64:216-223.

Smith, Philip W. 1955. Presumed hybridization of two species of box turtles. Chic. Acad. Sci., Nat. Hist. Misc. No. 146.

Waters, Joseph H. 1964. Subspecific intergradation in the Nantucket Island, Massachusetts, population of the turtle *Chrysemys picta*. Copeia 1964:550-553.

Speciation in Vascular Plants

The basic principles of the species concept are the same in the plant kingdom as in the animal kingdom. Separation of species is based on apparent morphological difference which presumably express genetic differences. In plants, however, environmental conditions have a more obvious effect upon morphological characteristics. Hybridization, polyploidy (multiplication of chromosome sets) and other phenomena complicate the picture. This exercise is designed to show how meristic characteristics are used in plant taxonomy, and how hybridization may be demonstrated. It will also demonstrate variation within a species.

Oaks (genus Quercus) occur in almost every part of our country. In most localities several species occur, and hybridization often occurs between certain species. Hybrids will normally indicate their origin by being intermediate between the parents, or by showing part of the characteristics of one parent and part of those of the other.

In the field, collect samples from eight oaks of each of two species. (Herbarium sheets may be used if field collection is impractical.) On the following data sheets, enter the data for each of your specimens. Then compare the two sets of data and examine for evidence of variation and hybridization. Do your samples agree perfectly with descriptions of the species in a standard botanical book? Do you find any intermediate specimens? Do any specimens from one species show characteristics typical of the other species? Do you think they may represent hybrids?

REFERENCES

Benson, Lyman. 1964. Plant classification. 2nd Ed. D. C. Heath and Co., Boston.

Muller, Cornelius H. 1952. Ecological control of hybridization in Quercus: a factor in the mechanism of evolution. Evol. 6:147-161.

DATA SHEET FOR COMPARISON OF OAK SAMPLES

STUDENT:______________________________ DATE:____________________

Characteristic	Sample No. 1	2	3	4	5	6	7	8
Leaf length								
Leaf width								
Number of lobes								
Shape of lobes								
Lobes bristle-tipped?								
Margin serrate or dentate								
Ventral surface color								
Dorsal surface color								
Ventral surface vesture								
Dorsal surface vesture								
Twig color								
Twig vesture								
Bud length								
Bud shape								
Bud color								
Bud vesture								
Fruit length								
Fruit width								
Ratio of cup lgth./acorn lgth.								
Cup shallow or deep								
Inside cup vesture								
Outside cup vesture								
Length of peduncle								

DATA SHEET FOR COMPARISON OF OAK SAMPLES

STUDENT:______________________________ DATE:__________________

Characteristic	Sample No. 1	2	3	4	5	6	7	8
Leaf length								
Leaf width								
Number of lobes								
Shape of lobes								
Lobes bristle-tipped?								
Margin serrate or dentate								
Ventral surface color								
Dorsal surface color								
Ventral surface vesture								
Dorsal surface vesture								
Twig color								
Twig vesture								
Bud length								
Bud shape								
Bud color								
Bud vesture								
Fruit length								
Fruit width								
Ratio of cup lgth./acorn lgth.								
Cup shallow or deep								
Inside cup vesture								
Outside cup vesture								
Length of peduncle								

Figure 1. DEGREE OF LOBING

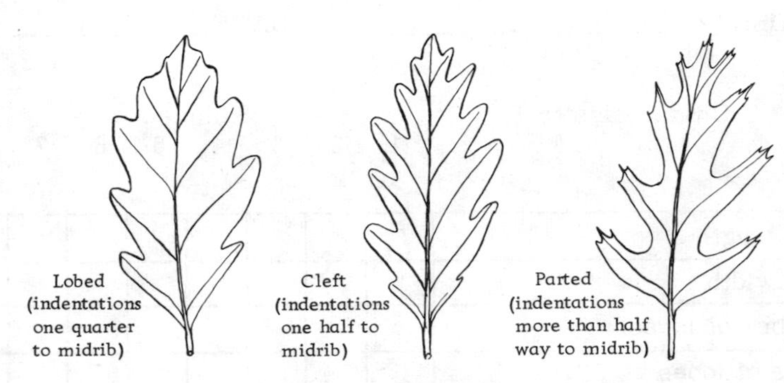

Figure 2. LEAF MARGINS

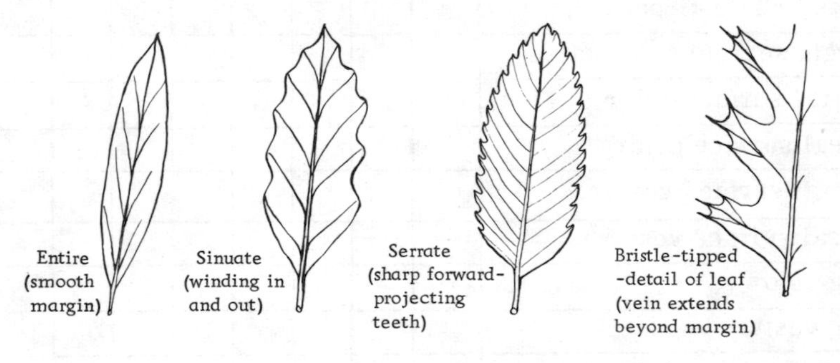

Figure 3. VESTURE

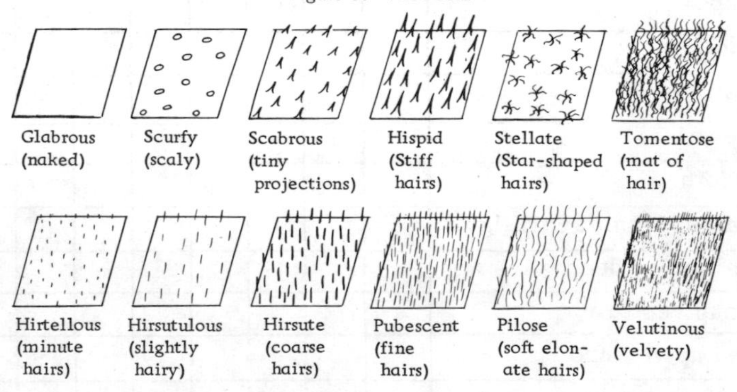

Section II
Some Field Techniques

How to Take Notes in the Field

A clipboard or spring-type holder is desirable for field use. A pencil can be attached with string, and should be used for all field notes, since ink may run when wet.

1. Take all notes on the spot. Do not trust your memory. It will fail you with alarming frequency, and this can only result in incomplete and inaccurate notes, which are of no value in a scientific study.
2. Make complete notes. You may have much that is unimportant, but your notes can be sorted out later to eliminate unnecessary material. It is better to have data which you do not use, than to need data which you do not have. At the time of taking notes, you cannot always decide what is important.
3. After you have completed the day's observations, stop to ask yourself if there is anything else you ought to check. Imagine yourself writing up results and look for items which are missing.
4. A picture, even a rough sketch, is worth a thousand words. Even though not artistic, a field sketch may help you to recall things you would not remember without such a sketch. It will also save time in many cases, for a picture can be used instead of a long description.
5. Cultivate the practice of observation. Learn to see things not ordinarily noticed. Write down what you see, avoiding interpretations based on inadequate information. With practice, you will learn to see the things which are important without conscious effort.
6. Be certain that observations are accurate. Identify all species carefully. If necessary, collect specimens for later identification when more time and adequate facilities are available. It is better to have no notes at all than to have notes filled with inaccuracies and errors.

REFERENCES

Cockrum, E. Lendell. 1962. Laboratory and field manual for introduction to mammalogy. Ronald Press Co., New York.
Hall, E. R. 1955. Handbook of mammals of Kansas. Univ. of Kansas, Mus. of Nat. Hist. Misc. Publ. No. 7, Lawrence.
Hickey, Joseph. 1963. A guide to bird-watching. Doubleday Anchor Books (reprint edition), New York.
Kelker, George H. 1958. Research methods for the beginner. Utah State Univ., Logan, U.
Mosby, Henry S., ed. 1960. Manual of game investigational techniques. The Wildlife Society, Ann Arbor, Mich.
Perry, Alfred E. 1965. Records please!! Turtox News 43(4):126.

Aids to the Field Biologist

Contributions to biology may still be made by the quiet observer with infinite patience and broad knowledge. For the most part, however, the field biologist has come to depend upon increasingly sophisticated instruments to aid him in his work. A number of books on these instruments and the techniques associated with their use have been written (see lists of references within this section). We can do no more here than to point out some of the more widely used instruments, and give a few hints as to their use.

BINOCULARS

For studying birds, binoculars are a necessity, and they are also used widely by mammalogists. Prism binoculars are most commonly in the range of 6 to 12 power, although higher powers are sometimes used on a tripod. Most users prefer 7 or 8 power, since the larger magnification necessitates heavier binoculars, making them harder to hold steady, and any movement is accentuated by the high magnification.

The user or prospective purchaser of binoculars needs to consider other things than magnification in choosing the right instrument to fit his needs. Perhaps most important is light transmission, especially if observations are to be made in dim light. This is usually indicated by the manufacturer as a numerical factor, but a good rule of thumb is that the diameter of the lens farthest from the eye should be at least five times the magnification. Thus binoculars are listed as 6 x 30, 7 x 35, etc. For those who will do considerable work at dusk or in dark places, 7 x 50 is the best choice; indeed, it is the preferred glass of a great many biologists, amateur and professional.

So-called featherweight binoculars cost a little extra, but are a good investment. These are made with light metals, and in sustained use are much easier on the muscles and easier to hold steady.

Most binoculars are made with a central focusing wheel, necessitating separate focus of one eyepiece. In some, the two eyepieces focus, and there is no central control. Some users prefer the former, but you can get used to either. Some older models, e.g. some military surplus glasses, may have individual focus, but this is probably not a major drawback.

REFERENCE

Reichert, Robert J., and Elsa Reichert. 1960. Binoculars and scopes. Chilton Co., Book Div., Philadelphia.

HAND LENSES

Many biologists, and botanists in particular, require these small magnifying glasses, usually of 10 to 14 power, for the study of small items. Triplet lenses of high optical quality are the most satisfactory models, but are of course more expensive. Good doublet lenses may be obtained from suppliers listed on p. 34 which would probably be satisfactory for the less demanding user.

PHOTOGRAPHIC EQUIPMENT

Virtually every field biologist finds it necessary, sooner or later, to use a camera in recording some aspects of his observations. So many different types and makes are available, however, that the beginner is likely to be thoroughly confused. We can do no more here than give a few general hints, but more detailed information can be found in the references listed below.

CAMERAS

Most field biologists are interested in an easily portable camera, which will take excellent pictures under a variety of light conditions, including both long-range photos and closeups. Probably the most common choice is a 35-mm. camera, which takes pictures about 1 x 1½ inches. Good 35-mm. cameras are of two types: single-lens reflex (hereafter referred to as SLR), and rangefinder models.

Many biologists feel that the SLR is the ideal field camera. The better models, selling from about $100 to $400, accept a variety of lenses and other accessories, take almost precisely what is seen in the finder (an advantage in closeup work), and are extremely versatile. Although the original investment is quite large, a good one is almost a lifetime investment and is in any case necessary for doing satisfactory photographic work.

Rangefinder models, led by the well-known Leica, Nikon, Contax and similar models, have long been favorites for their ruggedness and ease of use, combined with extreme versatility. Lower-priced models with less refinements may be found for as little as $25-50. They are easily and accurately focused, may be operated very rapidly, and, according to their proponents, are easier to focus than SLR's, as well as producing sharper negatives. These experts insist that the movement of the mirror which permits the SLR user to view the scene through the lens causes some movement and loss of sharpness. Equally worthy experts say that this is hogwash, and the beginner might perhaps be best advised to try each type for himself and make his own decision on the basis of which meets his needs and seems easiest to use.

Low priced cameras of many types are suitable for some uses (e.g., if enlargements are not required, sharpness is not critical, etc.), but a good camera is worth the investment. In many cases, a good second-hand camera may be found at a much lower price than a new one, and in excellent condition.

For extreme enlargements, a 35-mm. negative may not be adequate. The next-larger popular size is the 120, which takes pictures $2\frac{1}{4}$ x $2\frac{1}{4}$ (or in some models $2\frac{1}{4}$ x $3\frac{1}{4}$). The twin-lens reflex is the most popular of these cameras, although some good SLR cameras are manufactured in this size (e.g., Hasselblad, Bronica, Praktisix, etc.). Cameras of this type are an excellent compromise between the large press-type cameras and the smaller 35-mm. size. The more expensive models offer many accessories and are suitable for virtually all types of field use.

Large format cameras, press and view cameras which use film in sizes from $2\frac{1}{4}$ to $3\frac{1}{4}$ to 8 x 10, are preferred by some field workers because of the large negative. However, they are by nature bulky and require the use of a more expensive film. They have a complete line of lenses and accessories, and are especially suitable for landscape work, or wherever a large negative is desirable. Formerly, these cameras used cut film in sheets or film packs, but more modern models are adaptable to the use of roll film, Polaroid film, and extra long rolls of certain films.

FILMS

One of the major reasons for the overwhelming popularity of the smaller cameras is the development of extremely fine-grain films which will allow great enlargement of the small image. The number, variety and excellence of modern films presents the beginner with a confusing array. Basically, however, in the use of black and white film a simple rule is available: use the slowest film which will do the job you need to do. Films with a film speed rating of 400 or above are available if you need to take pictures by candlelight, but the negative quality will not be equal to that obtained with slower films. A medium-speed emulsion such as Kodak Plus-X Pan is a good choice for all-round use, but if maximum enlargement is desired and speed is not an object, a slower film such as Kodak Panatomic-X, Adox K-14, or any of several others will be the best choice. Complete lines of film are made by Eastman Kodak Company and Ansco in this country, and several brands of film are imported from Europe and Japan. In general, it is best for the beginner to learn to use one or two kinds of film well, rather than trying to sample a variety of brands whose differences are probably minor. The author's personal choice is a slow film, (ASA speed 16-40) developed in a slow-working compensating developer such as Agfa Rodinal, Tetanal Neofin Blue, or FR X-22. If you have your developing and printing done commercially, you lose much of the control over photographic quality, as well as much of the pleasure of photography. However, making your own is a time-consuming operation which may be impossible if large numbers of photographs are taken in the course of field work.

Each of the following references will give you a great deal more information than we can include here. For the novice, a beginner's guide such as the one from Eastman Kodak is desirable. As you pick up the basic facts, you will want to advance to one with more detailed information slanted toward your particular interests.

REFERENCES

Blaker, Alfred A. 1965. Photography for scientific publication. W. H. Freeman Co., San Francisco.

Eastman Kodak Co. 1959. The Kodak camera guide. Pocket Books Inc., New York.

Kinne, Russ. 1962. The complete book of nature photography. A. S. Barnes & Co., New York.

Linton, David. 1964. Photographing nature. Doubleday & Co., New York.

Shumway, Herbert D. 1956. Nature photography guide. Greenberg Publishers, New York.

MAPS

Geographical Survey maps are available for a great part of the country. They have a scale of 1 inch to 2000 feet (1:24,000), 1 inch to 1/2 mile (1:31,680), or 1 inch to 1 mile (1:62,500). Elevations are shown as contour lines, streams and water in blue. Many artifacts of man (houses, schools, bridges, etc.) are also indicated. Maps with forested areas in green are available on request. Free indices to the topographic maps of each state may be obtained by writing to the Geological Survey, Washington 25, D. C. From an index the map desired can be determined, then ordered, at 30 cents apiece. For the shores of the Great Lakes and the canals of the lakes there is also a series of maps available from the U. S. Lake Survey, Corps of Engineers, 630 Federal Bldg., Detroit 26, Michigan. Aerial photographs of many regions are obtainable either from state conservation departments or local farm bureau agents. Also, write to U. S. Dept. of Agriculture, Commodity Stabilization Service, Performance and Aerial Photography Division, Washington 25, D. C. Soil maps are available for many areas, and may be obtained from the Bureau of Plant Industry, Soils and Agricultural Engineering, U. S. Dept. of Agriculture, Washington, D. C. A file of aerial photographs and soil maps is often available for consultation at the local office of the Soil Conservation Service.

BIOLOGICAL SUPPLY HOUSES

Equipment for the field biologist cannot usually be secured in the ordinary store. Many supply houses catering to the needs of the biologist have arisen in various parts of the country. The list given here, while by no means complete, includes the largest and best known, and should permit the biologist to find any except the most specialized equipment.

Aloe Scientific Co., 5655 Kingsbury, St. Louis 12, Mo.
Carolina Biological Supply Co., Burlington, N. C. 27216
Cambosco, 342 Western Ave., Boston, Mass. 02135
Clay-Adams Inc., 141 East 25th St., New York 10, N. Y.
General Biological Supply House, Inc., 8200 South Hoyne Ave., Chicago, Illinois 60620
National Biological Laboratories, Inc., P. O. Box 511, Vienna, Va.
Nucleonic Corporation of America, 196 Degraw St., Brooklyn 31, New York (specializes in atomic research materials)
Powell Laboratories, Gladstone, Oregon 97027
Quivira Specialties Co., 4204 W. 21st St., Topeka, Kansas

Science Materials Center, 59 Fourth Avenue, New York 3, N. Y. (specializes in science education materials)
Standard Scientific Supply Corp., 808 Broadway, New York 3, N. Y.
Ward's Natural Science Establishment Inc., P. O. Box 1712, Rochester, N. Y. 14603 or P. O. Box 1749, Monterey, Calif. 93942
W. M. Welch Mfg. Co., 1515 Sedgwick St., Chicago 10, Ill.
Will Corporation, New York 52, N. Y. (other offices in various eastern cities)

A few more specialized suppliers are listed below:
Animal Trap Co. of America, Lititz, Pa. (snap traps of various types, etc.)
Allcock Mfg. Co., Box 551, Ossining, N. Y. (live traps)
National Agricultural Supply Co., Fort Atkinson, Wis. (forestry supplies, etc.)
H. B. Sherman, P. O. Box 683, Deland, Fla. (patented live traps for small mammals)
Forestry Suppliers, Inc., 205 W. Rankin St., P. O. Box 8397, Jackson, Mississippi 39202

Collection of Biological Specimens

There are few biological problems involving field work which do not eventually require collection of specimens. Learning the techniques of collection, therefore, is one of the first steps in becoming a field biologist. In the following section, brief suggestions are given for the beginner, and for each group of organisms, references are given which will give additional information. The following list of more general references will also be useful.

REFERENCES

Anderson, R. M. 1960. Methods of collecting and preserving vertebrate animals. Bul. 69, Natl. Mus. Canada, Ottawa.
Miller, D. F., and G. W. Blaydes. 1962. Methods and materials for teaching biological sciences. 2nd Ed. McGraw-Hill Book Co., N. Y.
Morholt, Evelyn, Paul F. Brandwein and Alexander Joseph. 1958. A source book for the biological sciences. Harcourt, Brace and World, N. Y.
Needham, J. G., ed. 1937. Culture methods for invertebrate animals. Comstock Publ. Assoc., Ithaca, N. Y. (Reprinted, 1959, by Dover Publ., New York).)
Pray, Leon. 1943. Taxidermy. Macmillan Co., N. Y.

Smithsonian Institution. 1944. A field collector's manual in natural history. Smiths. Inst. Publ. No. 3766, Washington, D. C.

Wagstaffe, R. J., and J. H. Fidler. 1955. The preservation of natural history specimens. Vol. 1, The invertebrates. Philos. Lib., N. Y.

Collection and Preservation of Mammals

A license is needed to collect game and fur-bearing mammals, but small mammals, such as mice, shrews and bats, may be collected without a license. Most small mammals may be taken in ordinary snap-back mouse traps, which may be purchased at any store. Live traps of various types can be quite readily constructed from woven wire, milk bottles, or boxes, or can be purchased from one of the companies listed on page 35.

Since small mammals are totally unsuspicious, traps do not need to be concealed. In fields, runways can be located along the surface of the ground. Traps set across these runways will catch meadow mice and short-tailed shrews, and occasionally other species. Active runways can be recognized by fresh cuttings of grass, fresh droppings, and a well-worn appearance. In woodlands, runways are to be found under logs and stones, and deep in the leaf mold. Some species do not utilize runways, and must be taken by setting traps at random in suitable habitat. In trapping any habitat, one series of traps should be set in such a manner, to insure a more representative sample of the mammal population than could be taken in runways.

Peanut butter, walnut meats, bacon, oatmeal and apple are good baits, either alone or in various mixtures. A bait of peanut butter, Roman meal and bacon grease is durable, easy to use and attractive to small mammals.

Small mammals occur almost everywhere, even in vacant city lots and yards. Most can be caught readily in snap traps. Special techniques must be developed to take moles, pine mice and other subterranean forms. Bats are best collected from their roosts late at night or during the daytime. They can also be collected by shooting over water and along woodland edges, but fine (No. 12) shot must be used to avoid serious damage to the specimens.

After the mammals are taken, accurate records should be made for future use. Such a record should be kept in a journal in India ink, and should include all information about your collecting experience, as well as other biological information. A field catalog, in which data on each mammal are entered, should also be carefully kept. Here are recorded the animal's number, scientific name, sex, measurements, weight, date and locality taken, and remarks on molt, condition of reproductive organs, stomach contents, parasites, or any other pertinent data. This serves as a supplement to the information which is on the tag attached to the skin and skull, and in case the tag is illegible at some future date, your catalog will provide the necessary substitute. Data sheets such as those provided (page 43) may be used as a catalog, or in addition to the catalog.

In keeping notes, it is not possible to be too complete. Everything is important; weather, habitat, time of day taken, details of habits observed, burrows, food, speed of movement, description of young, peculiar pelages, albinism, notes on parasites, in short any information which can add to our knowledge of the species concerned. For many of our small mammals, this information is sadly lacking, and could readily be obtained by an amateur worker who is in a locality where he can obtain numerous specimens.

It is sometimes difficult to determine externally the sex of young mammals or those in which the males do not posess a scrotum. In these cases, the sex may be determined by observing the distance from anus to clitoris or penis. In males, this distance is relatively longer than it is in females. (fig. 4)

In examining a female for reproductive data, the breasts may be squeezed to determine if the animal was lactating. If embryos are present, they should be measured from their crown to rump (fig. 5), and a note made of the number in each horn of the uterus. If there has been a recent partus, placental scars may remain. These are best observed by removing the reproductive tract intact, and placing it in water in a Syracuse watch glass or Petri dish and examining with a dissection microscope. The scars will appear as dark round spots. It is also sometimes possible to count the corpora lutea of each ovary. These appear as relatively large yellow bumps on the ovaries.

The examination of males is somewhat simpler. The testes of immature specimens of some species such as squirrels are located just under the skin between the vent and navel region. More often, the testes will be found in the abdominal cavity or in the scrotum. The length of the testes not including the epididymis, should be taken. If there is any doubt as to whether the animal is an adult or not, the epididymis should be observed for active sperm.

Often, when collecting, more specimens will be taken than may be examined or be "put up" as skins. If possible, these specimens should be skeletonized after examination. If there is not even time for this, they should be placed in 10% formalin or frozen.

Figure 4. SEX DIFFERENCE IN SMALL MAMMALS.

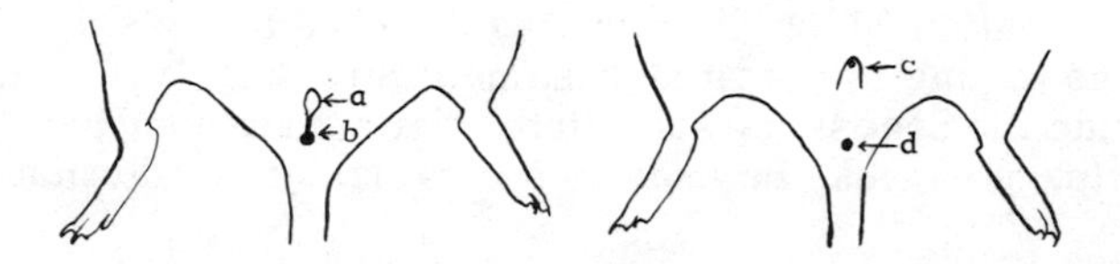

Left: relatively short distance between clitoris (a) and anus (b) in female

Right: relatively long distance between penis (c) and anus (d) in male

Figure 5. THE CROWN-RUMP (CR) MEASUREMENT OF EMBRYOS.

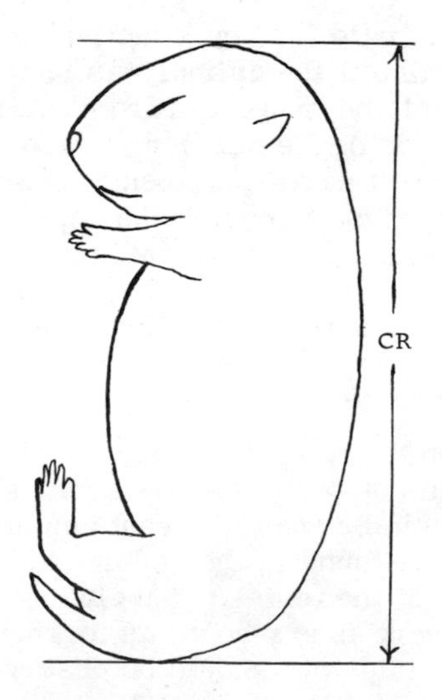

Figure 6. SKULL TAG 1" BY 2"

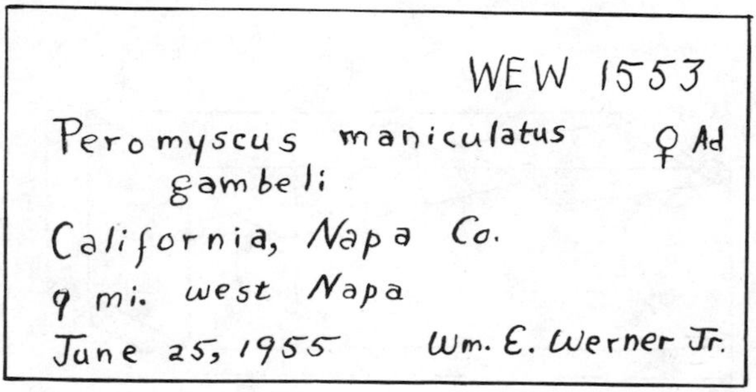

WEW 1553
Peromyscus maniculatus gambeli
♀ Ad
California, Napa Co.
9 mi. west Napa
June 25, 1955
Wm. E. Werner Jr.

Figure 7. MAMMAL SKIN TAG 3" BY 3/4"

WEW 1553
Peromyscus maniculatus gambeli
♀ Ad.
California, Napa Co.
9 miles west Napa
June 25, 1955
William E. Werner Jr.
3 placental scars, right horn of uterus
In canyon; hardwoods, redwood, mandrone, etc.
T.L. 156
T. 70
H.F. 20
Wt. 25g.

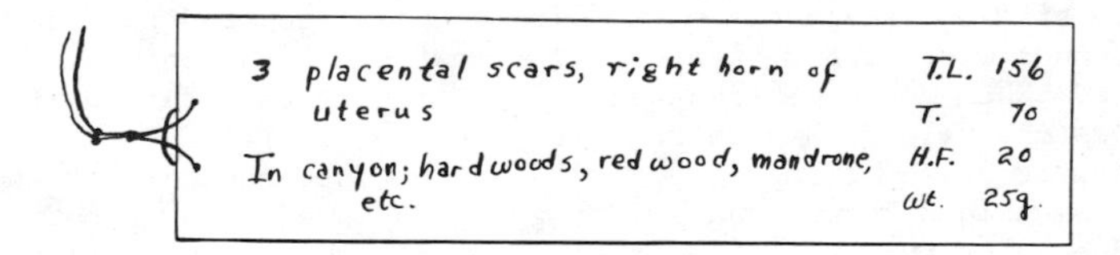

Upper: front side. Lower: reverse side.

Note that thread has two sets of knots, one near the tag, one ¼ inch out.

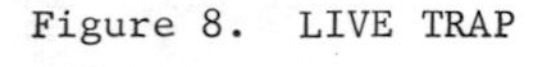

Figure 8. LIVE TRAP

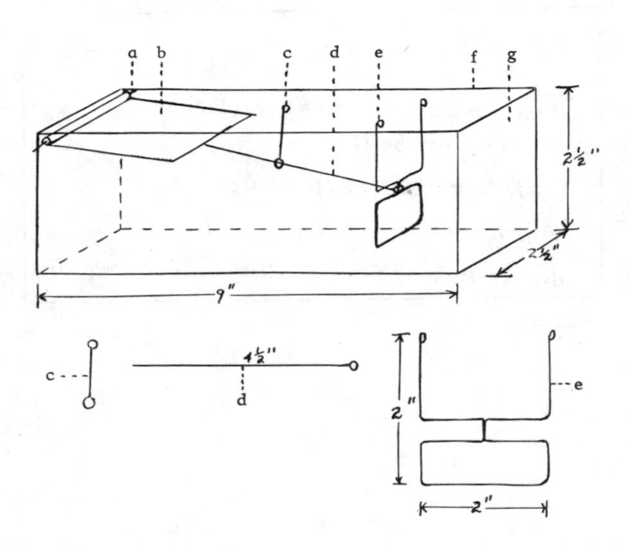

a. Wire pin to hold door, 3" long.
b. Door, aluminum, 2¼" wide by 3¼" long, with one end rolled around pin a.
c. Wire holder, one end around a cage wire, other end allows part d to slide freely through it. About 1½ - 1-3/4" long before bending.
d. Wire release, about 4½" long before bending. One end fits under door when trap is set, other end has loop loosely fitting around part e.
e. Wire trigger, shape as shown, with loops at top end to fit around a wire of the cage. Bottom must clear the bottom of the cage. When mouse pushes against the trigger to obtain bait, it moves the release enough to allow the door to fall.
f. Hardware cloth, ¼" mesh, 9" by 10", bent into rectangular shape of dimensions shown, and held together by bending over cut ends. Added security may be had by binding with wire or soldering.
g. Hardware cloth 2½ by 2½" to form back of trap. Attached as described for part f.

Some mice are able to open the door and escape. A lock can be devised consisting of a U-shaped loop, with the free ends bent around a wire of the cage between c and e, and with the U-shaped part above and against the door. It should be placed so that when the door falls, the lock also falls, bracing against the door about at floor level.

This is also true of specimens not suitable for skins and skulls or skeletons. Specimens preserved in this manner may be studied later when more time is available, and will provide information on reproduction, food habits, parasites, etc. This information should then be recorded in the journal.

When a mammal is taken in the field, it is wise to immediately place a small tag on it. This may bear information as to the exact location of its capture. The same tag may later be used as a temporary skull tag while the skull is being cleaned. If the specimen is not made into a skin and skull, this tag may be used as a specimen tag in the jar of preserved specimens.

Skulls should be cleaned, and then tagged (fig. 6) and stored. Colonies of dermestid beetles or meal worms may be kept for cleaning the skull. Large skulls may be cleaned by placing them in water. The water should be changed daily or more frequently, to slow down the bacterial action and minimize the odor. Preliminary cleaning procedure includes cutting off all excess flesh, including the tongue. The brain should be removed. If insects are used, the skull should be dried first. The final cleaning step is to bleach it overnight in hydrogen peroxide diluted 1:1 with water, and then to dry it quickly under a light. The number of the specimen should be written on the skull and each ramus of the lower jaw in India ink.

PREPARATION OF A MAMMAL STUDY SKIN

1. Make incision in skin of abdomen, either laterally from knee to knee, or from back of sternum to anus.
2. Remove skin along incision, until knees can be pushed up. Cut each hind leg at knee joint.
3. Remove flesh from leg-bones and pull back into place.
4. Skin around back until all is free except tail. Remove tail vertebrae by holding vertebrae with forceps and holding fingers in front of tail skin to prevent turning it inside out.
5. Strip skin over body to forelimbs. Cut forelimbs at knee.
6. Skin over back of head until skin begins to pull back at attachment of ears. Clip these close to skull.
7. Cut tissues connecting skin to skull around eye, being careful not to cut skin itself. Then skin around mouth, carefully cutting skin away from skull until free.
8. Cut off skull, tag it, and lay it aside.
9. Remove all excess flesh and fat, rubbing skin freely with borax.
10. Roll a piece of cotton to approximate the size of the body. After wrapping cotton around leg bones and replacing them in normal positions, turn skin inside out, so that fur is again outside. Then insert cotton body with forceps. Be sure to get it well into the head.

Figure 9. SMALL MAMMAL SKIN

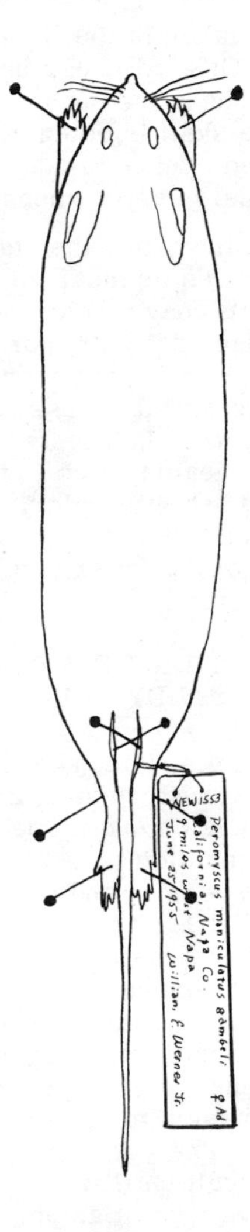

Note pins in forepaws and hindpaws (pads down), on either side of hind legs to hold legs against tail, and across tail to hold tail down. Label is on right hind leg above foot.

MAMMAL DATA SHEET

STUDENT:______________________________ DATE:______________

Species____ No.________

Sex________ Age_____ Condition______________________________

Weight__________ Measurement ______________________________

Date________________ Collector______________________________

Locality____

Habitat____

____

____

Reproductive data____

____

____

Molt data____

____

Stomach contents____

____

Remarks____

____

____

11. Wrap a thin layer of cotton around a wire about an inch longer than the tail, and insert it in the tail.
12. Pin to drying board, after sewing up mouth and ventral incision. Palms down, all feet close to body, fasten paper strip over head if head or ears tend to stand up.
13. Make out tag as follows: on front, your initials and the collection number of the specimen, scientific name, sex, locality, date, name of collector. On back, measurements, weight, remarks. (See fig. 7) Measurements taken, in millimeters, are total length, hind foot, and tail. Other measurements taken on bats are length of ear, and length of forearm.
14. Specimens must be kept in tight boxes in an atmosphere of paradichlorbenzene.

These directions will be much easier to follow if the student has had an opportunity to watch a demonstration of the correct technique. Certain "tricks of the trade" cannot be explained in a brief outline, but may be learned very quickly by watching an expert. If time permits, the instructor should give such a demonstration as part of the period or periods to be spent on learning to preserve animals.

REFERENCES

Beer, James. 1964. Bait preferences of some small mammals. J. Mammal. 45:632-634.

Cockrum, E. L. 1962. Laboratory and field manual for introduction to mammalogy. Ronald Press Co., New York.

Constantine, Denny G. 1958. An automatic bat-collecting device. J. Wildl. Mgmt. 22:17-22.

Manville, Richard H. 1949. Techniques for the capture and marking of mammals. J. Mammal. 30:27-33.

Collection and Preservation of Birds

A license from both State and Federal governments is required for collection or possession of bird specimens. It is usually not worthwhile for the amateur biologist to secure such licenses. However, some non-migratory birds are not protected by law, and these may often be picked up dead on the road. Therefore, it is well to know how to make a bird skin, so that these may be used.

The process of preparing a bird skin differs in a few details from that required for mammal skinning. These few differences will be stressed in the following account.

1. Make slit in skin from mid-breast to anus. Work skin over legs, cut them off, and work skin over back in the same manner as with mammals. Cut off tail close to body, leaving it attached only to skin.
2. Remove skin up to wings and cut wings close to body.
3. Skin up over head as in mammals, but leave skull in place. Make three cuts, one between mandibles which cuts through interorbital septum; one up each side of back of skull inside mandibles, then connect all three cuts with lateral cut across back of skull and into interorbital septum. Remove back of skull, brains, eyes, and tongue.
4. Clean flesh from leg and wing bones. Wrap leg bones with cotton and replace as in mammals. No cotton is needed on wing bones except in large birds.
5. Turn skin right-side out. Tie two humeri at inner ends, about 1/4 inch apart on small birds, up to 1 inch on crow-size birds.
6. Insert cotton balls into eye sockets. Wrap body cotton on sharp-pointed stick about as long as bird, and insert pointed end in upper mandible from inside the skin.
7. Sew up the abdominal slit, adding a triangular piece of cotton for additional stuffing if needed. Tie legs crossed to stick, smooth feathers naturally, and wrap in thin layer of cotton. It may be removed for further arranging after a few hours; then the skin should be left to dry for four days to a week.
8. Make out tag as for mammal, but no measurements need to be made, except weight. Skins must be kept in a tight container with paradichlorbenzene crystals, which must be periodically replaced.

As with mammals, a demonstration of making a bird skin will be a great help to the student. Bird skinning requires a bit more practice than does mammal skinning, but when well learned it can be done as quickly and easily. The skin of birds is, of course, more delicate, and care must be taken to avoid tearing it. Feathers, however, will cover a multitude of errors, so a specimen should not be abandoned because of a few minor difficulties. Even badly damaged specimens may be made presentable if enough care is taken.

A data sheet similar to that for mammals (p. 43) may be used in a collector's catalog for birds, with minor changes.

REFERENCES

Blake, E. R. 1949. Preserving birds for study. Fieldiana Tech. No. 7. Chicago Nat. Hist. Mus., Chicago.

Chapin, James P. 1946. The preparation of birds for study. Sci. Guide No. 58. Am. Mus. of Nat. Hist., N. Y.

See ALSO p. 224-225.

Collection and Preservation of Reptiles and Amphibia (Herptiles)

Among the vertebrates, the amphibians and reptiles are perhaps the most elusive and thus poorly known. The amphibians, frogs, toads and salamanders, are for the most part easily taken at their breeding ponds. Certain species, however, have a very brief and unpredictable breeding season (e.g., the spadefoot toad, which occurs over much of the country). Others, notably the Plethodontid salamanders, do not breed in the water. Some live underground most of the time, and indeed a few salamanders are known only from caves or deep wells.

Except for the obvious opportunity offered by the gathering of these animals at their breeding sites, there are few dependable ways of getting adequate samples. Some of the non-aquatic salamanders may be collected by overturning logs, stones and debris in suitable habitat. Frogs and toads may be caught with the aid of a flashlight as they seek food on rainy nights. Sometimes they may be seen crossing roads in considerable numbers. Perhaps the most important point is to know where and when to look. Study of one of the books listed at the end of this section will well repay the time spent.

Reptiles are most abundant in the south and southwestern parts of the country, but they may be found throughout the country. Snakes, lizards and turtles are the groups in which the average collector is interested. (Alligators and crocodiles are, of course, reptiles, but few except professional collectors are interested in them).

The time-honored, if arduous, method of catching snakes and lizards is to overturn stones and logs in the correct habitat. In this work, care must be taken to reach over the stone or other object and lift it from the far side. Thus if you uncover a poisonous snake (however unlikely that may be) your legs will be protected from a strike. Anyone who wishes to collect snakes should first familiarize himself thoroughly with any poisonous species which may occur in his collecting area. Snakes frequently may be taken while crossing roads at night or early in the morning.

Lizards, many of which are active during daylight, may be killed with a .22 rifle charged with dust shot. Some collectors have successfully used a slingshot, while others employ a blowgun made from a 6-foot length of 3/4 inch aluminum tubing. Darts are made of corks cut down to fit snugly without tapering. Needles are then inserted into the corks, blunt end first.

In any case, the most important factor in successful collecting is a knowledge of the organisms and their habits. This fact cannot be stressed too strongly.

Turtles are often picked up on land, especially the terrestrial and widespread box turtles. Aquatic turtles may be collected in a trap of chicken wire attached to floating logs. Bait such as bread or vegetables is placed in the middle of the enclosure. Once the turtles enter the trap they are unable to climb out.

Amphibians are easily killed by placing them in a 1% solution of chloretone, or in a weak solution of alcohol. This leaves them relaxed, and they may then be arranged in the desired position in a wax-bottomed dissecting pan and hardened with 10% formalin. Reptiles may be killed by ether (<u>not</u> chloroform), by injection of a lethal chemical, or if no other means is available, by drowning. After death, larger reptiles should be injected with formalin, and permanent storage should be in 10% formalin. Amphibians likewise should be stored in formalin after hardening. Ethyl alcohol, (70%) is also a satisfactory permanent storage medium, but for most small collections it is less convenient than formalin. Color retention may be prolonged by storage in a dark place.

REFERENCES

Banta, B. H. 1937. A simple trap for collecting desert reptiles. Herpetologica 13:174-176.

Bragg, Arthur N. 1949. Field preservation of amphibian eggs and larvae. Turtox News 27:262-263.

Conant, Roger. 1951. Collecting lizards at night under bridges. Copeia 1951:79.

Dorgan, Lucas M., and William H. Stickel. 1949. An experiment with snake trapping. Copeia 1949:264-268.

Gordon, Robert E., and Donald W. Tinkle. 1960. A technique for collecting reptile eggs. Texas J. Sci. 12:14-16.

Lagler, Karl F. 1943. Methods of collecting freshwater turtles. Copeia 1943:21-25.

Neill, Wilfred T. 1950. How to preserve reptiles and amphibians for scientific study. Ross Allen Reptile Inst., Silver Springs, Florida.

Tinkle, Donald W. 1956. Blowguns for reptile sampling. Southwestern Naturalist 1:133-134.

See ALSO p. 224.

HERPTILE DATA SHEET

STUDENT:______________________ DATE:______________

Species ______________________________ No.__________

Sex ________________ Collector____________________

Date ____________________ Locality______________________

____

Habitat ____

____

____

Measurements____

Weight______________ Reproductive data__________________

____

Remarks (color when alive, behavior, etc.)________________

____

____

____

____

____

____

____

____

____

____

____

Collection and Preservation of Fishes

Fishes are usually collected with nets of various types and sizes. The ten-foot to twenty-foot seine is generally most suitable. In large streams or lakes, a bag seine or a gill net may be used. The use of hook and line may be desirable to take certain game species not easily netted. Bait fishes and other small species may be collected in minnow traps. There are many state and local regulations in regard to taking fishes. Be certain that you know these regulations, and that you have the necessary licenses or permits.

Fishes may be killed by placing them in a solution of chloretone for a few minutes. As soon as they are relaxed, they may be placed into 10% formalin for hardening. Large specimens should be injected with the preservative. Some curators prefer 70% alcohol as a permanent preservative.

Specimens should be placed in jars, and data in regard to their collection should be placed in each jar. Data should be written with engrossing ink on good quality paper, and should include data, locality, collector, and number of the collection. Additional data should be written on data sheets, such as the sample on the following page. Data labels should always be placed inside the specimen jar, never stuck on the outside. Glued labels are easily lost, and specimens without data are of limited value.

REFERENCES

Hubbs, Carl, and Karl Lagler. 1958. Fishes of the Great Lakes Region. Cranbook Inst. of Sci. Bull. No. 26, Bloomfield Hills, Mich.

Scott, W. B. 1954. Freshwater fishes of eastern Canada. Univ. of Toronto Press. Toronto, Ontario, Canada. Univ. of Toronto Press.

See ALSO p. 223-224.

Collection and Preservation of Aquatic Invertebrates

Protozoa and other microscopic forms of aquatic life may be obtained by collecting water samples from various levels. Surface scum, bottom ooze, and scrapings from plants are often especially rich in microscopic animal life. Another method of collecting organisms floating in the water, is by the use of a special fine-meshed net. These nets (plankton nets) are towed through the water at various desired levels under the surface behind a boat, or are placed in a stream, allowing the current to sweep organisms into them. Small sized nets may be thrown from shore and pulled in by an attached line. Since these are usually studied alive, preservation is not necessary. If it is desired to keep material for further study, cultures may be made, or specimens can be kept in alcohol or formalin.

Sponges, hydras and flatworms are usually found attached to underwater vegetation, stones and debris. They may be taken by putting such material in a white pan filled with water, and removing individual organisms with forceps or a small brush. Ten percent formalin or 70% alcohol will serve as a preservative.

Leeches, nematodes and other aquatic worms may be seen in the water and can be collected with a dip-net. Leeches are also often taken from the shells of turtles. Crayfish may be caught by hand, or in a dip-net. Snails may be collected readily by hand if the water is clear, or by drawing a net through aquatic vegetation. For large or deep bodies of water, a throw-net with a long rope attached is helpful. A net placed downstream in rapid water will catch many organisms if the rocks or bottom above are stirred. The use of a glass-bottomed pail or diving mask permits selectivity in the collecting of these organisms.

Preservation may be accomplished by placing the specimens in 70% alcohol. Adequate data should accompany each collection, including at least the date, locality, and collector. Further data in regard to the habitat, activity of the organisms, etc. are desirable.

Care should be taken in aquatic collecting to observe all legal regulations. Seines are forbidden in some streams, and there may be other local or state laws which govern collecting of this sort. The collector should always check with local law-enforcement agencies.

REFERENCES

Lauff, G. H., et al. 1961. A method for sorting bottom fauna samples by elutriation. Limnol. and Oceanogr. 6:462-466.

Needham, John G., and Paul Needham. 1962. A guide to the study of freshwater biology. Holden-Day Inc., San Francisco.

Ryther, J. H., C. S. Yentsch, and G. H. Kauff. 1959. Sources of limnological and oceanographic apparatus and supplies. Limn. and Oceanog. 4:357-365.

Shelford, V. E., and Samuel Eddy. 1929. Methods for the study of stream communities. Ecology 10:382-391.

FISH DATA SHEET

STUDENT:________________________________ DATE:____________________

State______________________County________________Township_________

Drainage________________________ Locality__________________________

_____________________________________ Coll. No.____________________

Water depth______________Width___________Current______________

Bottom__________________Vegetation____________________________

Temperature: Air_____Water__________Weather_________________

Depth of Collection__________Method of Collection________________

Collector________________________________ Date____________________

Remarks____

____

____

Species collected:

Collection and Preservation of Insects

Terrestrial and flying insects are usually captured by means of an insect net, used to beat the bushes, sweep the vegetation, or catch insects in flight. These nets are of several types, each designed for use in one of these ways. These nets and other materials mentioned below may be obtained from biological supply houses (see page 34). Some species may be collected by turning stones, removing bark from dead or dying trees, or sifting leaf mold. A highly specialized group of insects may be found in fungus or carrion. Some species like sweets, and may be attracted by splashing molasses on trees.

As any city-dweller is aware, insects are attracted to street lights. This proclivity may be exploited by hanging a Coleman lantern or an electric bulb in a well-wooded area, at an elevation such that it can be seen for a long distance. A white sheet or cloth should be suspended nearby as a surface on which insects may alight.

Insects are usually killed by placing them in a cyanide jar, which contains a mixture of sawdust and potassium cyanide, held in place by a thin layer of plaster of Paris. These jars are dangerous to have where children may get them, are difficult to dispose of, and potassium cyanide is hard to secure in some circumstances. A safe and satisfactory substitute can readily be made. Place a half inch of sawdust or other absorbent material in a jar which has a tight lid. Cover the sawdust with a half-inch or less of plaster of Paris. Just before going afield, pour into the jar a teaspoonful of ethyl acetate, an inexpensive chemical which can be obtained at any drugstore. Ethyl acetate must be added before each trip, since it evaporates and loses its potency in a day or two. Methyl acetate (banana oil) also is quite satisfactory.

Most insects are mounted on insect pins, available from biological supply houses. These pins come in various sizes to suit the size of the insect. Very small insects may be mounted on small triangular pieces of paper with a bit of glue, or on special tiny pins called minuten nadeln. The smallest insects (fleas, lice, etc.) are usually mounted on slides so that they may be studied with a microscope.

Most insects are pinned through the right side of the thorax. Beetles are pinned through the front of the right wing cover. True bugs (Hemiptera) are pinned through a triangular area between the front ends of the wings (the scutellum). Butterflies and moths must be dried with the wings spread and pinned in place. Special spreading boards are available, but a substitute can be made by cutting slits in a corrugated cardboard box. Slits of varying sizes can be made to accommodate the bodies of the insects, and the wings are spread at the sides. They should be allowed to dry for at least a week before being removed.

Dragonflies and damselflies often lose wings and bodies when kept on pins, so that they are often kept in plastic or cellophane envelopes.

Insect labels of very small size (1/4 x 1/2 inch) are placed on the pin below the specimen. These include locality, date and name of collector. A second label giving the name of the insect may be added later, and other labels giving host plant of insect, habitat, or other data may be added as needed.

Soft-bodied insects such as silverfish, as well as larvae and pupae and some aquatic insects, are best preserved in 70% alcohol to which a small amount of glycerin has been added. Collection of aquatic insects may be accomplished by the means discussed under aquatic invertebrates, as well as by specialized bottom-scrapers and other instruments which may be obtained from the biological supply houses.

REFERENCES

Borror, D. J., and D. M. DeLong. 1964. An introduction to the study of insects. 2nd Ed. Rinehart & Co., New York.

Oldroyd, H. 1958. Collecting, preserving and studying insects. Macmillan Co., New York.

Oman, P. W., and A. D. Cushman. 1948. Collection and preservation of insects. U. S. D. A. Misc. Publ. No. 60, Washington, D. C.

Peterson, A. M. 1953. A manual of entomological techniques. 7th Ed. Published by the author, Columbus, Ohio.

Collection and Preservation of Botanical Specimens

Plants are much easier to collect than animals, since they do not move around, for the most part, and can be readily taken. The only equipment needed in the field is a carrying case, commonly known as a vasculum, a trowel to dig up herbs, and a knife to remove samples of woody plants.

In the case of small plants, the entire plant should be collected, and flowering or spore-bearing specimens should be taken if available. Average sized, typical specimens should be taken, and if possible both flowers and fruit of seed plants should be collected. Fertile and sterile fronds of ferns, if both occur, are needed. The roots are often necessary for identification, and leaves from various parts of the plant are desirable.

If the plant is too large to fit on the herbarium sheet, or in the vasculum, selected parts should be taken to show all parts of the plant: roots, basal leaves, center section of stem, flowers and fruits, etc. In the case of woody plants which are much too large for this procedure, leaves, twigs, flowers and fruits are needed. For such plants, good photographs of parts too large to collect may be of great value.

Before going into the field, fold wet newspapers and place them in the bottom of the vasculum, to keep the plants moist until they can be pressed. Some very delicate ferns and aquatic plants need to be pressed in the field, as soon as they are taken.

A plant press may be secured from a biological supply house, or may be built by riveting 1 x 1/4 inch hardwood slats into a lattice about 12 x 18 inches. Two of these form the sides of the press. Newspapers or large sheets of paper are used to enclose the plant, and these are surrounded by soft builder's felt or blotter paper. Corrugated cardboard sheets may be inserted to insure better aeration. The arrangement of the plant in the folded sheet is important, since this is the form in which it will be permanently fixed. It should be carefully prepared, with all necessary characteristics showing clearly. Different leaves should be arranged so that both top and bottom of leaves are shown. If the plant is too long for the sheet, it may be folded zigzag and pruned to fit. Care must be taken not to remove plant sections which are necessary for identification. A knowledge of the group concerned is the best guide to securing the proper type of material for preservation. Leaves, stems and flowers are always needed, while roots, seeds and fruit are also sometimes needed for identification.

After the plant is ready to press, lay out the sheet of paper or newspaper, and place the plant on it. Arrange the plant as desired, and carefully fold the paper over it. Holding it in place with one hand, place it on a piece of felt, and place another piece of felt on top of the newspaper, removing the hand as you do so. Then place the entire pile on the bottom half of a plant press, add a piece of corrugated cardboard, and you are ready to begin with another plant. When the plants are all in the press, place the other half of the press on top and lash tightly with straps. It can then be placed in an oven or drying room or near a radiator for drying.

In about four days to a week, the plants will be ready to mount. Plants may be fastened to the herbarium sheet with gummed linen tape, or may be glued. If glue is used, a large flat dish or a glass plate covered with glue is placed alongside the pile of plants. As the plant is removed from the press, it is dropped briefly onto the glue, lifted out, and placed on the herbarium sheet in the desired position. If tape is used, the plant must be taped at several points to insure adequate attachment and to prevent future breakage. For plants which are to be handled much, glue is preferable.

A label about 2 x 4 inches in size is usually placed in the lower right corner of the sheet. This label should be glued firmly in place, and should contain the following data: date, place, collector, habitat, elevation, soil type, scientific and common name of the plant if known, and any other information which may be of importance.

Lower plants are kept in a different manner. Algae are best preserved in a solution known as FAA solution. This solution is prepared by mixing 5 parts commercial formalin, 5 parts glacial acetic acid, and 90 parts 50% grain alcohol. A small amount of copper sulfate is sometimes added to improve color retention. Fungi and lichens are often dried, while mosses may be dried or kept in liquid. Fruits and seeds of various kinds and flowers for dissection are also preserved in liquid. Plants may also be quick-frozen for perfect preservation.

Marine algae may be preserved by floating out specimens in pans of salt water (fresh water will bleach out the red pigments especially). Herbarium paper is then slipped carefully under the specimen, and its position arranged. The herbarium paper is then slowly lifted up, without disturbing the position of the algae, until it is out of the water. The specimen and herbarium sheet are both placed in a plant press with blotters on each side, followed by corrugated cardboard. The procedure is repeated for each specimen. Natural glue on the algae will cause it to stick to the herbarium paper. To prevent the algae from sticking to the blotters, a piece of wax paper must be placed over the specimen.

Botanical specimens which are mounted on cards are kept in special moth-proof cabinets. If such cabinets are not available, specimens may be kept in cabinets well saturated with paradichlorbenzene crystals, and fumigated occasionally with carbon disulfide. Herbarium sheets are usually filed in manila folders, either in alphabetical order by genus or in phylogenetic order.

REFERENCES

Benson, Lyman. 1964. Plant classification. 2nd Ed. D. C. Heath Co., Boston.

Johnston, I. M. The preparation of botanical specimens for the herbarium. Reprinted and distributed by Cambosco Sci. Co., 342 Western Ave., Boston, Mass.

Pool, Raymond, J. 1941. Flowers and flowering plants. McGraw-Hill Book Co., N. Y.

See ALSO pps. 218-221.

Animal Signs

In making a survey of animals in a given area, it is desirable to use every possible clue in detecting the presence and abundance of each species. Sometimes it is not possible or desirable to kill or trap animals, nor are all species easily observed. For example, many of the accepted records of the mountain lion in the northeast in recent years are based on tracks or other signs. In census work in quadrats, all mammals and most birds would probably be scared away before they could be observed, yet they are an important part of the community. If we are to evaluate these species in an ecological study, we will have to make use of some of the indications which they often leave behind to show that they have been present.

Various types of animal signs may be used as clues to occurrence and numbers, including food cuttings or caches, claw marks, footprints, nests, tunnels, scats, food remains, pellets, etc. Some of the more widespread and easily observed of these signs are given here, while a more complete treatment of the subject may be found in the references listed below.

Food cuttings and caches: Meadow mice (*Microtus* sp.) frequently cut grass into 3/4 to 1 inch lengths, which are scattered through the runways or in small piles. Lemming mice (*Synaptomys* sp.) have neat piles of grass stems of uniform length along their runways. The grass cuttings of meadow jumping mice (*Zapus* sp.) are longer, 2 to 3 inches in length.

Browsed shrubbery is a clue to the identity of the browser. Rabbits cut the branches off cleanly, while deer break and pull them off, leaving a ragged end. Trees cut by beavers appear almost like trees cut with an axe, but tooth marks can be seen.

Food stores are maintained by many rodents. Stores of nuts and seeds under logs in woods are often placed there by deermice (*Peromyscus* sp.) Underground stores of seeds, roots, tubers and nuts may be made by chipmunks, pine mice, meadow mice, and others. Heaps of pine cone remains around a log or stump often indicate the presence of a red squirrel (*Tamiasciurus* sp.).

Claw Marks: Probably the easiest of claw marks to recognize are those made by bears on standing trees. Scratches made by squirrels on dead trees are often a clue to the location of a nest. Raccoons and other species often scratch on trees, but such marks are difficult to identify accurately.

Footprints: In warm seasons the muddy shores of streams and ponds will reveal the tracks of many wild creatures, including raccoons, opossums, mink, weasels, crows, herons and many others. In winter, the snow provides a perfect opportunity to study animal tracks in colder climates. A light fall of fresh snow on top of older harder snow makes ideal tracking conditions, and tracks of such tiny mammals as shrews and deermice are interlaced with the larger tracks of skunks, foxes, and other mammals. Winter birds too leave characteristic tracks, but these are harder to identify. Tunnels under the snow may show the activity of meadow mice and shrews.

Nests: Tree nests of squirrels and deermice are easily observed and recognized. Meadow mice build hollow balls of grasses in grassy meadows on the surface. Underground nests of many speies are equally typical, though seldom seen. Bird nests are most easily located in winter, and are usually identifiable. Shrubby areas undergoing secondary succession are especially productive of accessible bird nests.

Tunnels: The numerous tunnels of mammals may be felt underfoot as one walks over areas of abundance. Shrews, moles, meadow mice, pine mice and other species may participate in the construction and use of these runways, but food cuttings, as indicated above, may offer a clue as to the occupants. Mole tunnels are characterized by raised ridges of earth in many cases. Gophers leave mounds of earth periodically along their burrow systems, giving a characteristic appearance. Burrows of larger mammals, such as woodchucks, foxes and skunks may be recognized by hair, odor or remains of food around the burrow entrance.

Food remains: The feeding habits of many predatory animals are so unique that food remains may indicate the predator responsible for the kill. The following examples cover some common and widespread species.

Horned owl: Large victims are disemboweled and eaten on the ground. Smaller victims are carried to a roosting tree. Neck meat of birds is eaten, and meat is removed without disarticulation of the skeleton. Large (about 3 inches long) oval pellets of hair, bones and feathers are disgorged, often in piles around a favorite roosting tree.

Sharp-shinned and Cooper's hawks: These bird-eating hawks pluck most of the feathers from the victim before feeding begins. Prey is carried or dragged under nearby cover, head and limbs are removed, and the viscera and then the flesh are eaten. Neck vertebrae may be picked clean without disarticulation. Piles of feathers and skeletal remains are often clues to such feasts and chalky fecal matter may be found in the vicinity of such kills.

Crows and Jays: These birds often eat eggs, piercing them with the bill. Crows usually remove eggs from the nest and eat them elsewhere. Eggs which have hatched normally will usually be cracked cleanly along a line, while those eaten by predators are likely to have one or two smaller openings through the shell.

Skunks: Shallow digging in open fields often show where skunks have dug for grubs. Eggs are often destroyed by skunks, being opened along the long axis and eaten at the site. Some of the shell may also be eaten.

House cat: In some areas the house cat may be an important predator of small prey. Remains of cat kills are usually scattered over a wide area. Tooth marks are visible on the bones, and some feathers or fur may be eaten along with the meat.

Foxes: Large prey is not eaten at one time, and considerable parts are often found at the site of the kill. The legs of birds are left and are often flexed. The bones of the victim will bear tooth marks, but these differ from cats in that all of the marks will be about the same size. In cats, the larger upper canine teeth leave larger marks.

Scats: The fecal droppings of mammals may be used to note the presence of the animal doing the defecation, and in the case of large predatory mammals, may contain the bones and fur or feathers of prey species. Some of the common scats to be found in the field are noted here:

Field mice: approximately 2 mm. long, green-brown, in runway "toilets"

Shrews: approximately 2 mm. long, dark brown, in runways

Bats: 1-2 mm. long, brown to black, on floor of barns or caves, often best clue to location of roosting spots

Weasel: slender, twisted, about 1 inch long

Cottontail rabbit: spherical, about 8-10 mm. in diameter

Snowshoe hare: spherical, about 12-15 mm. in diameter, frequently with fibrous content

Deer: cylindrical to ovate, 10-25 mm. long, 8-10 mm. in diameter; in winter oval, pellet-like

Porcupine: spherical to oblong, 15-25 mm., smooth surface often with fibrous content

Beaver: spherical to irregular, 20-25 mm. in diameter, appearing like pressed sawdust

Skunk: cylindrical, 15-50 mm. long, frequently with remains of insects

Fox: cylindrical, 2-6 inches long, often with fur, feathers, and bones

REFERENCES

Einarsen, Arthur S. 1956. Determination of some predator species by field signs. Oregon State College, Corvallis, Ore.
Headstrom, Richard. 1949. Bird's nests: a field guide. Ives-Washburn Inc., New York, N. Y.
Headstrom, Richard. 1951. Bird's nests of the west: a field guide. Ives-Washburn, Inc., New York, N. Y.
Murie, Olaus. 1954. Field guide to animal tracks. Houghton-Mifflin Co., Boston, Mass.
Stains, Howard. 1962. Game biology and game management: a laboratory manual. Burgess Publ. Co., Minneapolis, Minn., pp. 9-36.

Stomach Content Analysis

One method of determining the food habits of a species is to examine many stomachs of the animals in question. Stomach analysis is also of value in regional survey studies, for occasionally stomachs of predators may disclose the presence of species not obtained by ordinary collecting methods. This method may be used with any of the vertebrates, although analysis is easier and frequently more fruitful with predators than with herbivores. Occasionally, intestinal contents will provide additional information, especially if the food items are vertebrates. Frogs provide a convenient and easily obtainable vertebrate for this study, and their stomach contents are usually readily identifiable.

Obtain several specimens and remove their stomachs by cutting the esophagus, and duodenum near the pyloric valve. Place each stomach in a suitable container of water, such as a Syracuse watch glass or Petri dish. Remove the contents of the stomach, and identify the items. The more detailed the identification is, the better, even though at a later date results may be pooled into larger taxonomic categories.

Two types of data should be recorded for each item found in a stomach: its identity, and the approximate per cent of the total volume of the stomach contents that this kind of food composes. The per cent volume will be only an estimate, but it may be improved by using a grid, such as that of 1/4 inch graph paper, under the glass. For <u>each</u> kind of food found, use a separate column, and indicate the per cent volume. Each entry will also indicate an occurrence. An example follows:

Specimen Number	Diptera	Phalangida	Mites	Coleoptera	Spider
1.	25%	75%			
2.	40%	50%	10%		
3.	50%			40%	10%
4.	35%			65%	
5.	30%			60%	10%

When all the stomachs have been examined, the data should be summarized. To find the per cent frequency occurrence that any one item appeared in all the stomachs of the sample, add up the number of stomachs in which the food item appeared, and divide by the number of stomachs examined. Thus if Diptera were found in only 5 stomachs, and 34 stomachs were examined, the per cent frequency occurrence of Diptera would be 14.7% (5/34). To find the per cent volume that any one item constituted in the whole sample, add up all the per cent volumes of this item in all the stomachs, and then divide by the total of the per cent volumes of all the food items. (This latter figure should be 100 x the number of stomachs.) If Diptera formed 25% of the food in one stomach, 40% in a second, 50% in a third, 35% in a fourth, and 30% in a fifth, the sum of per cent volumes of Diptera would be 180. This sum is then divided by 100 x the number of stomachs examined (Grand per cent volume). In our example, the grand per cent volume is 3400, and the per cent volume of Diptera occurring in this sample of frogs is therefore 5.3% (180/3400). Each student should tabulate his data as illustrated below.

Since each student's sample will be small, the instructor may also want to pool the data of the whole class to illustrate effects of sample size on statistical data of this kind.

Stomach Contents of 34 Rana palustris taken 2 miles North of Alexandria Bay, St. Lawrence River, Jefferson County, New York, July 10-21, 1953.

Food Item	Per cent Frequency Occurrence	Per cent Volume
Coleoptera	41.2	24.8
Spider	33.4	9.0
Snail	26.5	12.8
Lepidoptera larva	23.5	9.5
Orthoptera	20.6	13.5
Hymenoptera	20.6	6.5
Diptera	14.7	5.3
Hemiptera	11.7	2.5
Grass, pebble	11.7	1.5
Undetermined arthropods	8.8	4.1
Phalangida	5.9	3.7
Undetermined insects	5.9	2.6
Homoptera	5.9	1.7
Trichoptera larva	5.9	1.1
Sow bugs	5.9	0.7
Collembola	2.9	0.3
Mites	2.9	0.3
		99.7

REFERENCES

Hamilton, W. J., Jr. 1948. The food and feeding behavior of the green frog, Rana clamitans Latreille, in N. Y. S. Copeia (1948) No. 3:204-207.

Haskell, William L. 1959. Diet of the Mississippi threadfin shad, Dorosoma petense atachafalayae, in Arizona. Copeia (1959) No. 4:298-307.

Klimstra, W. D. 1959. Foods of the racer Coluber constrictor, in Southern Illinois. Copeia (1959) No. 3:210-214.

Martin, A. C. 1949. Procedure in wildlife food studies. Wildlife Leaflet 325. Fish and Wildlife Service, Washington, D. C.

__________, A. L. Nelson, and H. S. Zim. 1951. American wildlife and plants. McGraw-Hill Book Co., N. Y.

Williams, Olwen. 1959. Food habits of the deer mouse. J. of Mammal. 40:415-419.

STOMACH CONTENT ANALYSIS

DATA SHEET

STUDENT:________________________ DATE:______________

Name of specimens:______________________________________

Number:______________ Location:__________________________

Date of capture:____

Food Items

Specimen number						
Totals						
Per cent frequency						
Per cent volume						

Ecological Methods

The science of ecology has emphasized the principle of measurement in field biology. This includes measurements of factors of the physical environment, as well as various measurements of statistics involving organisms. Some of the more important chemical, physical and biological measurements are described here. For many of the chemical ones, it is often wise to practice them first in the laboratory before using them in the field.

OXYGEN CONCENTRATION IN WATER

Glass bottles with tight-fitting glass stoppers and having a capacity of 250 cc. are used. Make certain all glassware used is clean. Special water samplers are available to secure water without introducing atmospheric oxygen. If a sampler is not available, the water should be obtained in such a manner as to avoid splashing or bubbling the water. Siphons are often of use in such a procedure. Temperature of the water should also be recorded at the time and place the water is collected. If water is then transferred from the sampler to the bottle, the water should be allowed to overflow the 250 cc. bottle 2-3 times to flush out atmospheric oxygen. When the stopper is replaced, no air bubble should remain.

Now proceed to analyze the water as follows: (This is known as the Rideal-Stewart modified Winkler test.) For steps #1-6 the reagents should be added quickly and the bottle restoppered to prevent oxygenation from the atmosphere.

1. To the water sample add:
   a glass bead to aid in mixing
   .7 cc concentrated H_2SO_4
   1.0 cc. $KMnO_4$ solution
2. Shake well. A pale violet to pink color should appear and persist. If it does not, add another cc. of $KMnO_4$ solution. After the color is established, allow the sample to stand for at least 40 minutes.
3. Add 1.0 cc. potassium oxalate solution. Let stand until the color disappears.
4. Add: 1.0 cc. manganous sulfate solution
   3.0 cc. hydroxide sodium iodide solution
5. Shake. A yellow precipitate will form. Allow this to partially settle and then shake again.

6. Add .5 cc. of concentrated H_2SO_4. The precipitate should dissolve. If it does not, add another .5 cc. of the acid. The yellowish color remaining represents the iodine which has replaced the dissolved oxygen. At this point analysis may be suspended for some time, allowing the titration to be carried out in the laboratory.
7. Measure out 100 cc. of the water sample and titrate with the sodium thiosulfate solution. MAKE SURE YOU READ THE LEVEL OF THE SODIUM THIOSULFATE SOLUTION BEFORE YOU START TITRATING. WRITE IT DOWN. Titrate until a very pale yellow color is reached.
8. Add 2 cc. of starch solution. The sample will turn blue.
9. Continue titrating until the sample becomes clear. The clearness should persist under agitation.
10. Calculate the amount of sodium thiosulfate used, in cc., and multiply by 4/5. This will give roughly the parts per million of dissolved oxygen in the sample of water. More exactly, PPM dissolved O_2 =

$$\frac{800 \text{ x cc. thiosulfate used x normality of thiosulfate}}{\text{cc. of sample titrated}}$$

REAGENTS

All water used should be distilled. Salts are usually dissolved in small quantities of water first, then diluted.

Potassium permanganate solution: 6.32 gm. $KMnO_4$ in 1 liter H_2O.

Potassium oxalate solution: 20 gm. $K_2C_2O_4 \cdot H_2O$ dissolved in water. Add 4 gm. NaOH and dilute to 1 liter with water.

Manganous sulfate solution: 480 gm. $MnSO_4 \cdot 4\ H_2O$ to 1 liter of water.

Hydroxide-sodium iodide solution: 500 gm. NaOH and 135 gm. NaI to 1 liter of water.

Sodium thiosulfate: (N/10): 24.82 gm. $NaS_2SO_3 \cdot 5\ H_2O$ to 1 liter of water. Use cooled boiled water. Add 5 cc. of chloroform. When this is to be used, dilute to N/100 by adding 9 parts water to 1 part of this solution.

Starch solution: 3 gm. potato starch, ground with H_2O. Place in 500 cc. freshly boiled water. Allow to stand overnight, then use only the clear fluid.

FREE CO_2 CONCENTRATION

Obtain a water sample and with it fill a Nessler tube to the 100 cc. mark. Be careful not to splash or agitate the water, since carbon dioxide will easily come out of solution. Proceed immediately to analyze the sample as follows:

1. Put in 10 drops of phenolthalein solution. (5 g. phenolthalein in 1 liter of 50% alcohol, neutralized with N/50 NaOH)
2. Titrate with N/44 NaOH. Be sure to write down the level of NaOH before starting the titration.
3. The end point is reached when a pink color appears for a few seconds under agitation. Do not titrate until a permanent pink color forms.
4. Read the burette after titration, calculate the amount of NaOH used.
5. Multiply this amount by 10. This will give the amount of free CO_2 in parts per million that was in the sample.

HYDROGEN ION CONCENTRATION

An approximate value of hydrogen ion concentration of water, expressed in pH, may be obtained by using special indicator papers. These papers are merely dipped in the water, and the resultant color is compared to a color standard. For testing soil pH, a small sample of soil may be placed in 5 cc. of distilled water and shaken well. Allow the mixture to settle and then use the indicator paper. The soil should be tested at ground level and then in the sub-surface. This type of test is not precise, and gives only an indication of the pH of soil or water. It may be quite inaccurate under certain circumstances, such as in highly organic soils.

TEMPERATURE

Special devices including reversing thermometers, electric resistance thermometers and bathythermographs are available to determine temperature at any depth of water. An easy way to accomplish this is to lower a maximum-minimum thermometer into the water, which will record the coldest or warmest temperature reached. For shallow water, air and soil, however, ordinary mercury thermometers serve the field biologist well. Temperatures of water should be taken at varying depths and in different locations from the surface to the bottom. Temperature of the air should be taken at the surface of the water or soil, and at different levels in the air. Temperature of the soil should be taken just under the leaf litter and at varying depths.

RELATIVE HUMIDITY

The usual device to record relative humidity is the psychrometer, which contains a wet-bulb and a dry bulb thermometer. The wet bulb thermometer shows a lower temperature due to evaporation of water from a wick around its bulb. Since the rate of evaporation depends on moisture already in the air and on air temperature, the differences in temperature between the two thermometers can be translated into relative humidity, if the barometric pressure is known. Tables are available for this translation (see Oosting, 1956). For ecological work these thermometers are usually mounted so that they may be whirled in the atmosphere, and hence are called sling psychrometers. Another type involves the rapid intake of air through a chute passing it over the bulbs. The latter device enables measurement of relative humidity close to the ground or water, and in small cavities.

REFERENCES

Anonymous. 1955. Standard methods for the examination of water, sewage and industrial wastes. Am. Public Health Assn., Inc., N. Y. 522 pp.

Huffaker, C. B. 1942. Vegetational correlations with vapor pressure deficit and relative humidity. Am. Midland Naturalist. 38:486-500.

Lagler, Karl F. 1956. Freshwater fishery biology. Wm. Brown Co., Dubuque, Iowa.

Lund, J. W. G., and J. F. Talling. 1957. Botanical limnological methods with special reference to the algae. Botan. Rev. 23:489-583.

Thornthwaite, C. Warren. 1940. Atmospheric moisture in relation to ecological problems. Ecology 21:17-28.

Waring, R. H., and J. Major. 1964. Some vegetation of the California Redwood Region in relation to gradients of moisture, nutrients, light and temperature. Ecol. Monogr. 34:167-215.

Welch, Paul S. 1948. Limnological methods. Blakiston Div., McGraw-Hill Book Co., N. Y.

PLANT POPULATION ANALYSIS

The numbers and kinds of plants in any environment are usually determined by counting all specimens in a given sample area. The two kinds of areas used most frequently are squares and strips, the former called quadrats, the latter transects.

If many quadrats are to be established, selection of their sites should be at random. They may be set up at regular intervals in a line or in a grid. If only a few quadrats are to be used, they should be selected so as to include a representative sample of plants in so far as possible.

Each quadrat is a square established by using a rope or chain of a certain length, and then marking out a square with it. The type of vegetation to be analyzed determines the size of the quadrat. Thus a square with 10 meter sides is often used for trees, 5 meters for shrubs, and 1 meter for herbs. The smaller squares may be within the larger, and thus the number of shrubs in the 5 meter square multiplied by 4 will give the projected number of shrubs in the 10 meter square, and the number of herbs in the 1 meter square multiplied by 100 will give the projected number of herbs in the 10 meter square. Counting is sometimes implemented by the removal of the plants as they are noted.

Where the vegetation is varied in type, so that selecting typical quadrats is difficult, a transect may be used. This is a line or strip of narrow width extending through a community. All the plants in the strip or along the line may be enumerated. A scheme similar to that employed in quadrats may be utilized to simplify the counting of trees, shrubs, and herbs.

REFERENCES

Cain, S. A. and G. M. deOliveira Castro. 1959. Manual of vegetation analysis. Harper & Bros., N. Y.
Cooper, Charles F. 1963. An evaluation of variable plot sampling in shrub and herbaceous vegetation. Ecology 44:565-569.
Cragg, J. B., Ed. 1962. Advances in ecological research. Academic Press, N. Y. 203 pp.
Curtis, J. T. 1956. Plant ecology workbook. Burgess Pub. Co., Minneapolis.
Gates, Frank C. 1949. Field manual of plant ecology. McGraw-Hill Book Co., N. Y.
Grieg-Smith, P. 1964. Quantitative plant ecology. 2nd Ed. Academic Press, N. Y.
Leach, William. 1957. Plant ecology. John Wiley & Sons Inc., N. Y.
Lindsey, Alton A., James D. Barton, Jr., and S. R. Miles. 1958. Field efficiencies of forest sampling methods. Ecology 39: 428-444.
Neebe, David J., and Stephen G. Boyce. 1959. A rapid method of establishing permanent sample plots. J. Forestry. 57:507.
Oosting, Henry J. 1956. The study of plant communities. W. H. Freeman Co., San Francisco.
Phillips, Edwin A. 1959. Methods of vegetation study. Henry Holt & Co., N. Y.
Platt, R. B., and J. Griffiths. 1964. Environmental measurement and interpretation. Reinhold Publ. Co., N. Y. 235 pp.
Techniques and methods of measuring understory vegetation. (Symposium) 1959. Southern Forest Exper. Sta. and U. S. Dept. of Agr.

ANIMAL POPULATION ANALYSIS

Due to the motility of animals, censusing them is more difficult. Flying insects may be captured by sweeping vegetation with insect nets. This is best accomplished before any other sampling is done which might disturb the insects. Insects so caught may be counted, but of course this may not present an accurate picture of the insect populations. (See pp. 171-172.)

After a quadrat has been set up, animals present on the vegetation and ground may be noted and counted. When the plant census is finished, the soil may be spaded to a depth of 3-6 inches, and sifted with a coarse screen for animals. Again, this will give only a partial picture of the animals present, for many small insects and invertebrates such as nematodes may escape observation.

In forest situations, leaf litter may conceal many small animals. The litter may be removed from a certain portion of a quadrat and placed in a Berlese funnel. This is a container with a large funnel at the bottom leading to a collecting bottle. A sieve separates the container and the funnel. A light bulb at the top of the container dries out the litter, driving the animals through the sieve and down the funnel. Invertebrates may then be quantitatively and qualitatively analyzed.

Methods of determining the size of populations of vertebrates are described under the techniques of marking wild animals (page 73) and in population studies of birds (page 157) mammals (page 150), and amphibians (page 162).

REFERENCES

Andrewartha, H. G. and L. C. Birch. 1954. The distribution and abundance of animals. Univ. of Chicago Press, Chicago.

Cole, Lamont C. 1946. A theory for analyzing contagiously distributed populations. Ecology 27:329-341.

Elton, Charles S. and Richard S. Miller. 1954. The ecological survey of animal communities: with a practical system of classifying habitats by structural features. J. Animal.Ecology 42: 460-496.

Mosby, Henry. Ed. 1960. Manual of game investigational techniques. The Wildlife Society. Virginia Polytechnic Inst., Blacksburg, Va.

Platt, Robert B., and J. Griffiths. 1964. Environmental measurement and interpretation. Reinhold Publ. Co., N. Y. 235 pp.

Warren, Katherine B. Ed. 1957. Population studies; animal ecology and demography. Cold Spring Harbor symposia on Quant. Biol. vol. 22. Biol. Lab. Cold Spring Harbor, N. Y.

Techniques for Marking Wild Animals for Study

The oldest technique of marking animals for future identification is probably branding. For scientific purposes, bird-banding, or ringing, as it is known in Europe, has been in use for more than a century. The same technique has been found useful in banding bats, the band being placed about the forearm, loosely enough to slide without injuring the membrane. For small mammals, toe clipping may be used. A system of numbering is set up, so that an animal with a particular toe or toes clipped represents a given number. The toe is cut off just behind the nail with sharp scissors. The damage to the animal is apparently not significant if not more than two or three toes are clipped, and the mark is permanent. The only drawback to this system is that small mammals sometimes lose toes accidentally, and such individuals will cause a certain amount of confusion.

Various techniques besides branding have been developed for marking large mammals. Patches of hair clipped off in a pattern will last until the next molt. Various dyes have been used on both mammals and birds, but these too are temporary. Ear tags are perhaps the most widely used marks for permanent recognition, but these, of course, require capture of the animal, while the other marks mentioned can be observed from a distance. Tattooing likewise is permanent but detectable only close at hand.

The most modern technique of marking is that of placing a small pellet of radioactive material in a protective band, and attaching this to the animal. Animals marked in this way can be followed with a Geiger counter, or can be located at a later date without trapping. This method has been used successfully with mammals as small as the field mouse.

The most successful way of marking amphibians is toe clipping, as described above for mammals. (See also p. 165.) Metal clips have been placed in the lower jaw of some large species with fair success, but some fall out and others cause infection.

Snakes are marked by clipping scales on the ventral surface of the tail. The scales are numbered from the anus backward, and since they occur in two rows, a great many numbers are available. Turtles may be marked with paint for a short term study. Holes bored in the edge of the carapace in a particular pattern may serve for permanent recognition. Lizards may be toe-clipped in the manner described above for small mammals, and for amphibians.

Several methods of marking are in use for fishes. Small tags are sometimes placed in the jaw or other suitable part of the body. Portions of fins may be clipped in a manner similar to toe-clipping, but care must be taken not to clip enough to injure the fish or reduce its changes for survival, and of course the number of fins available is small so that large series cannot be marked in this manner.

Recovery of marked animals has been used as an index to the total population, by the use of a simple ratio known as the Lincoln-Petersen Index. This formula is:

$$P : M :: p : m$$

when P equals the total population, M equals the total number of marked individuals, p equals the total number of individuals taken at the second collection, and m equals the number of marked individuals taken at the second collection. Thus if 36 individuals were marked, and at a later date 40 individuals were found, of which 10 were marked, the total population would be:

$$P \text{ equals } \frac{36 \times 40}{10} \text{ or } \frac{1440}{10} \text{ or } 144.$$

REFERENCES

Symposium: Uses of marking animals in ecological studies. Ecology 37(4):665-689. 1956.

Introduction: Angus M. Woodbury. Page 665.

The marking of fish. Wm. E. Ricker, Pages 665-674.

Marking amphibians and reptiles. A. M. Woodbury. Pages 670-674.

Marking birds for scientific purposes. Clarence Cottam. Pages 675-681.

Marking of mammals; standard methods and new developments. Richard D. Taber. Pages 681-685.

Labeling animals with radioisotopes. Robert C. Pendleton. Pages 686-689.

Manville, Richard H. 1949. Techniques for the capture and marking of mammals. J. Mammal. 39:27-33.

See ALSO section on biotelemetry, below.

Biotelemetry

The rapid advances in instrumentation in recent years, which have produced such fruitful results in molecular biology, physiology and other laboratory disciplines, have not been without effect on field biology. Miniaturization of electronic instruments has made it possible to attach instruments of various types to living animals, or even to insert them into larger animals, and thus transmit information from the organism and/or its environment to a biologist at some distance away.

Probably the oldest, and certainly the most widely used application of biotelemetry consists of the attachment of a radio sending device so that the animal can be tracked. Studies of home range of animals, movements of animals within a community, even tracking of migrating birds, may be accomplished with this technique. Some enterprising biologists follow birds with airplanes, to stay within range of the radio.

More complex equipment can be used to measure changes in body temperature, rate of heartbeat, and other physiological characteristics, while the animal continues its activity in a presumably normal manner. The advance in this type of work which have evolved through the space program have found their way into more earth-bound uses of many kinds.

Animal behavior, too, has benefitted from these tools and techniques. Correlation of behavior patterns with electrical potentials in the nervous system has shown promise of leading to better understanding of the control of behavior and of the manner in which the nervous system exercises this control. Changes in the pattern of broadcasts can tell the listener what kind of activity a bird or mammal is engaged in, thus providing a means of monitoring activity patterns without actually following the animal.

The actual electronic requirements of such instruments are beyond the scope of this manual, but may be found by referring to the papers and books listed below. Any project in which movements, behavior and/or physiological changes in an unrestrained animal are desiderata, provides an opportunity for the application of this new technique. Miniaturized transmitters are now available which weigh less than half a gram and occupy a fraction of a cubic centimeter. The power source is the component which is most difficult to reduce, but studies in progress indicate that in some cases it may be possible to draw on the body energy of the animal and thus eliminate the battery.

REFERENCES

Adams, Lowell. 1965. Progress in ecological telemetry. Bioscience 15(2):83-86.

Cochran, W. W., and R. D. Lord, Jr. 1963. A radio-tracking system for wild animals. J. Wildl. Mgmt. 27:9-24.

Gold, D. C., and W. J. Perkins. 1959. A miniature electro-encephalograph telemeter system. Electron. Eng. 31:337-339.

Mackay, R. S. 1964. A progress report on telemetry from inside the body. Biomed. Sci. Instr. 2:275-292.

Sanderson, G. C., and B. C. Sanderson. 1964. Radio-tracking rats in Malaysia - a preliminary study. J. Wildl. Mgmt. 28:752-768.
Slater, L. E., ed. 1963. Bio-telemetry. Pergamon Press, N. Y.

In addition to the above, BioScience 15(2), February, 1965 includes a dozen papers on modern aspects of biotelemetry, with an admirable bibliography compiled by Lowell Adams. Interested students should read this summary, which covers the whole field of biotelemetry briefly but well.

Several recent books have discussed instrumentation in biology, and though most of this material is of use in laboratory work, some field techniques are included. The book by Platt and Griffiths, however, is specifically oriented toward ecological studies.

Kay, R. H. 1964. Experimental biology: measurement and analysis. Reinhold Publ. Corp., New York.
Newman, D. W. 1964. Instrumental methods of experimental biology. Macmillan Co., New York.
Platt, Robert T., and John F. Griffiths. 1964. Environmental measurement and interpretation. Reinhold Publ. Corp., New York.
VanNorman, Richard W. 1963. Experimental biology. Prentice-Hall Inc., Englewood Cliffs, New Jersey.

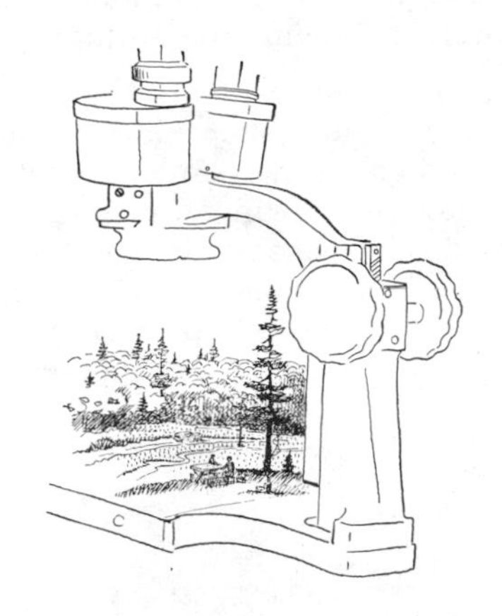

Section III

Terrestrial Communities and Succession

Introduction

One of the basic concepts of ecology is that of the organic community of interdependent and interacting organisms. One of the most interesting aspects of natural history is that procession of events which we term succession. By treating the various stages in this succession as communities, it is possible to emphasize two broad concepts with a single field trip. The stages may be treated in a series of trips, if time allows, or a trip may be planned so as to include several stages and thus several communities, at the discretion of the instructor.

Whenever an area of land is denuded of vegetation, and then left to itself, it undergoes a regular series of events which result at last in return to a type of vegetation which is self-perpetuating and relatively stable, unless major climatic changes occur. A final stage of this kind is called a climax community. Its organic makeup will vary greatly with variations in climate, topography, moisture, soil, etc. Although the process of succession is a continuous one, with no real natural breaks, we can subdivide it, for convenience, into four stages, which can be considered individually.

In regions where the climax is desert or grassland, succession also occurs, but there are fewer major stages of development. For these areas, only the climax stage is included here.

REFERENCES

Allee, W. C., A. E. Emerson, O. Park, and K. P. Schmidt. 1949. Principles of animal ecology. W. B. Saunders Co., Phila., Penn.

Benton, Allen H. and William E. Werner, Jr. 1965. Field biology and ecology. McGraw-Hill Book Co., N. Y.

Braun, Emma Lucy. 1950. The deciduous forest of eastern North America. McGraw-Hill Book Co., N. Y.

Clements, Frederic E. 1928. Plant succession and indicators. H. W. Wilson Co., N. Y.

Daubenmire, R. F. 1947. Plants and environment. John Wiley & Sons, N. Y.
Dice, Lee R. 1943. The biotic provinces of North America. Univ. of Mich. Press, Ann Arbor.
____________1952. Natural communities. Univ. of Mich. Press, Ann Arbor,
Egler, Frank E. 1951. A commentary on American plant ecology, based on the textbooks of 1947-1949. Ecology 32:673-695.
Elton, Charles. 1947. Animal ecology. Third Ed. The Macmillan Co., N. Y.
Golley, F. B. 1960. Energy dynamics of a food chain of an old-field community. Ecol. Monogr. 30:187-206.
Hanson, Herbert C. and Ethan D. Churchill. 1961. The plant community. Reinhold Publishing Co., N. Y.
Just, Theodore, ed. 1939. Plant and animal communities. Amer. Midland Naturalist 21:1-225.
Kendeigh, S. Charles. 1961. Animal ecology. Prentice-Hall, Englewood Cliffs, N. J.
Kittrege, Joseph. 1948. Forest influences. McGraw-Hill Book Co., Inc.
Odum, Eugene P. 1959. Fundamentals of ecology. W. B. Saunders Co., Phila.
____________1960. Organic production and turnover in old field succession. Ecology 41:34-49.
Olson, Jerry. 1963. Energy storage and the balance of producers and decomposers in ecological systems. Ecology 44:322-331.
Oosting, Henry J. 1956. The study of plant communities. W. H. Freeman Co., San Francisco.
Shelford, Victor E. 1963. The ecology of North America. Univ. of Illinois Press, Urbana.
Whittaker, R. H. 1953. A consideration of climax theory; the climax as a population pattern. Ecol. Monogr. 23:41-78.
____________1958. Recent evolution of ecological concepts in relation to the eastern forests of North America. In Fifty Years of Botany, Ed. by W. C. Steere. PP. 340-358. McGraw-Hill Book Co., N. Y.
For films and filmstrips on terrestrial communities, see pp. 228-229.

The Pioneer Weed Stage

The early stages of succession are marked by the dominance of grasses, weeds, and in the case of formerly cultivated fields and pastures, by the presence of many domesticated plants such as clovers, alfalfa, etc. This stage may be preceded, in completely denuded areas, by a moss-lichen stage, but where cultivated land has been abandoned this stage does not often occur.

The dominant plant groups in this community are the grasses (family Gramineae) and the composites (family Compositae) such as asters, goldenrods, and dandelions. Weeds of many other families may be present. In wetter areas, sedges and rushes may largely replace the grasses, and in very dry or impoverished soils certain hardy kinds of grasses may be about the only plants able to persist. If the land was previously forested and then cleared, woody species may persist from roots, and may continue to sprout for many years.

The dominant animal forms in point of numbers are invertebrates: spiders, insects, particularly of the orders Orthoptera (crickets and grasshoppers), Hemiptera, (the true bugs) and Hymenoptera (bees, wasps, and ants). On and in the soil, springtails (Collembola) may be abundant, and slugs and snails are found. Soil nematodes and other small soil organisms are also present. Among the vertebrates, birds are usually the most conspicuous, especially during the breeding season. Sparrows, meadowlarks, bobolinks, and such larger species as the pheasant, quail, and marsh hawk may nest here, while others may use it as a feeding place. Hawks and owls hunt small mammals, and insectivorous and seed-eating birds may also seek food around its edges at least. Mammals, though not always obvious, are usually numerous. Meadow mice (*Microtus*) and shrews (*Blarina*, *Cryptotis* and *Sorex*) may be present in large numbers. Other mammalian inhabitants include meadow jumping mice (*Zapus*), rabbits and woodchucks. Deer may feed here, and many predators, attracted by the smaller mammals, may include weasels, skunks, foxes, and domestic cats and dogs.

Reptiles and amphibians are not generally common at this stage, although certain species, such as the grass frog (*Rana pipiens*), garter snake (*Thamnophis* sp.) and green snakes (*Opheodrys* sp.) may be found. In those parts of the country where lizards are found, these early stages of succession are often well-populated with them. Those reptiles and amphibians which are present are all supported by the tremendous invertebrate population which exists here.

FIELD WORK IN THE PIONEER WEED COMMUNITY

Set up one or more quadrats, one meter square, and count accurately the plants within this space. Record data on the quadrat data sheets (page 99). If the field trip is to be of two or three hours duration set at least 50 snap traps in a small area upon arrival, and examine them for small mammals when leaving. Make collections of insects and other invertebrates, and record the relative abundance of different orders. Record all vertebrates seen, and seek for signs of some not actually seen. Such signs may consist of tracks in a muddy spot, droppings, nests of birds, feathers, hair, etc. If desired, the class may be divided into groups of four or five, each group being assigned to one of these projects. A group discussion can be held to summarize the results of the work of each group.

What environmental factors control plant distribution in this community? Is there any evidence of sprouting trees from a previous forest in the study area? Why may seedling trees not spring up at once in open fields near woodlands? Can animals ever control or markedly affect plant succession? What are the dominant plants and animals in the community? Is there a tendency for certain species of plants to occur in nearly pure colonies? Is this type of vegetation found as a climax anywhere in this country?

REFERENCES

Beckwith, S. L. 1954. Ecological succession on abandoned farmland and its relation to wildlife management. Ecol. Monogr. 22: 195-215.

Buss, Irven O. 1956. Plant succession on a sand plain, northwestern Wisconsin. Trans. Wisconsin Acad. Sci. 45:11-19.

Egler, Frank E. 1954. Vegetation science concepts. I. Initial floristic composition, a factor in old-field vegetation development. Vegetatio 4:412-417.

Evans, Francis C., and Eilif Dahl. 1955. The vegetational structure of an abandoned field in southeastern Michigan and its relation to environmental factors. Ecology 36:685-706.

Flaccus, Edward. 1959. Revegetation of landslides in the White Mountains of New Hampshire. Ecology 40:692-703.

Golley, Frank B. 1965. Structure and function of an old-field broom sedge community. Ecol. Monogr. 35:113-131.

Hale, Mason E., Jr. 1959. Studies on lichen growth rate and succession. Bull. Torrey Botan. Club. 86:94-129.

Hirth, Harold F. 1959. Small mammals in old field succession. Ecology 40:415-425.

Keever, Katherine. 1950. Causes of succession on old fields of Piedmont, North Carolina. Ecol. Monogr. 20:229-250.

Poogie, John J., Jr. 1963. Coastal pioneer plants and habitat in the Tampico Region, Mexico. Louisiana State Univ. Studies. Coastal Studies Ser. No. 6. State Univ. Press, Baton Rouge.

Tisdale, E. W. and M. Hironaka. 1964. Secondary succession in annual vegetation in southern Idaho. Ecology 44:810-812.

See ALSO references on pp. 77-78.
For identification manuals, see pp. 218-226.

The Shrub Stage

At an early time in the development of the weed stage, seeds of woody plants begin to sprout in the cool moist area around the bases of the herbs. These shrubs include some species which can survive direct sunlight and can compete successfully with the grasses and weeds which are still dominant. After a time the shrubs have grown higher than the herbaceous plants and have spread more or less throughout the area, although they may occur in clumps for the most part. They then become dominant, that is, they control the vegetation which may grow in company with them. Under their shade, sun-loving grasses and composites die, and mosses which had been growing around the bottoms of the plants spread and multiply. Also in their shade, seeds of certain trees begin to grow, so that this stage of development "sows the seed of its own destruction." Sometimes, however, stands of shrubs may be so dense as to exclude seedlings of trees, and this may arrest the progress of succession for 25 to 100 years at least.

The plants which make up the dominant growth in the shrub community may vary greatly from place to place. Meadowsweet (Spiraea sp.), dogwood (Cornus sp.), alder (Alnus sp.), arrowwood (Viburnum sp.), hawthorn (Crataegus sp.) and blackberry (Rubus sp.) are among the widespread forms which may occur in great numbers.

Because of the disappearance of much of the herbaceous vegetation, invertebrates are less abundant here than in the weed stage. As a result, those animals which feed on these invertebrates are also less common. A shrub community in its early stages, when grasses are still abundant, may harbor 20 pairs of birds per acre, while one consisting largely of shrubs, with relatively little herbaceous growth, will support hardly half that number. Mammal, reptile and amphibian populations are similarly reduced. Shrews are present, as are their subterranean relatives, the moles. Chipmunks may live on the seeds of the shrubs, sharing them with a few deermice (Peromyscus sp.). Several species of birds find this habitat to their liking, including thrashers, towhees, several species of warblers, and some sparrows. Among the reptiles, lizards and a few species of snakes occur. Toads are the only amphibians to be expected. The total number of animals per unit area in this type of vegetation is, however, considerable, and predators do not neglect its possibilities. The added cover may protect certain species from predation, but the lack of low ground cover prevents other species from living in such a habitat.

FIELD WORK IN THE SHRUB COMMUNITY

Since a quadrat one meter square would include only a few shrubs, set up several quadrats of five or ten meters square. Collect invertebrates from the surrounding area and list the orders found. If time permits, a trapline should be set up, although the scarcity of small mammals in this community would require an overnight trapping operation if success is expected. Look for animal signs, such as dust baths, burrows, runways, nests, etc. Turn over stones or other material under which organisms might hide. Dig up some shrubs and trees to ascertain whether they are sprouts or seedlings. If sprouts, can you tell how many times they have resprouted?

Compare the plant families and species here with those in the pioneer weed stage. How do total numbers compare? Does your collection of invertebrates show any significant variation in numbers or groups from that in the earlier stage? Are signs of vertebrate life as abundant here? What new species of vertebrates did you record? What are some of the factors which control the presence or absence of vertebrates here? What are the most important factors controlling the vegetation? Did you find tree seedlings under the shrubs? Of what species? How old is the largest shrub you can find? Which appear to be older, trees or shrubs? In what areas of the country do shrubs make up the climax vegetation?

REFERENCES

Aller, Alvin R. 1956. A taxonomic and ecologic study of the flora of Monument Peak, Oregon. Am. Midland Naturalist 56:454-472.

Fautin, Reed W. 1946. Biotic communities of the northern desert shrub biome in western Utah. Ecol. Monogr. 16(4):251-310.

Hanson, Herbert C. 1953. Vegetation types in northwestern Alaska and comparisons with communities in other Arctic regions. Ecology 64:111-140.

Mueggler, Walter F. 1965. Ecology of seral shrub communities in the cedar-hemlock zone of northern Idaho. Ecol. Monogr. 35: 165-185.

Niering, William A., and Frank E. Egler. 1955. A shrub community of *Viburnum lentago*, stable for twenty-five years. Ecol. 36: 356-360.

Patric, James H. and Ted L. Hanes. 1964. Chaparral succession in a San Gabriel Mountain area of California. Ecology: 45:353-360.

See ALSO references on pp. 77-78.
For identification manuals, see pp. 218-226.

The Intermediate Tree Stage

Among the shrubs, tree seedlings thrive, and eventually push their way above the shrubs. If they belong to certain fast-growing and sun-tolerant species, they may shoot up several feet in a year. In a few years such species will completely outstrip the shrubs, which die out as they are shaded by trees. These trees include poplars (Populus sp.) birches (Betula sp.), cherry (Prunus sp.) or any of several other species. Although such species grow rapidly, they have a relatively short life, so that the intermediate tree stage is often temporary. In some instances, however, trees such as pines may be the first trees, and they may dominate a community for a century or more before being replaced by other tree species.

The advancing shade has killed not only the sun-loving plants of the weed stage, but also most of the light-tolerant shrubs. Some, such as sumac (Rhus sp.) and berries (Rubus sp.) may persist, at least around the edges. Shade-loving plants which are typical of the forest do not immediately find their way into the trees, so that for a few years the vegetation on the forest floor may be sparse. Mosses and lichens are abundant, and several species of ferns occur. As woody plants die and decay, fungi become abundant. The shape of things to come may be seen in the presence of young seedlings of longer-lived trees such as oaks, maples and hickories.

Animal life is not markedly different from that of the shrub stage, although some groups are more abundant. Spiders, slugs, snails, millipedes and sowbugs may be found. Among the insects, flies (Diptera) and wasps and bees (Hymenoptera) are most obvious, although the soil surface harbors springtails (Collembola) and other forms. Wood-boring beetles of various families find homes in the dying trees.

A few species of warblers and vireos nest in the small trees or on the ground, but birds continue to be present in relatively small numbers. Mammals which can live on the invertebrates, or on fungi and plant seeds, are able to thrive. These may include species of deermice and shrews and moles. Other species may use the habitat for resting, cover, or food-hunting, but it is not favorable as a home for most species.

Reptiles and amphibians are restricted to a few species which find suitable habitat. Salamanders and frogs which are typical of climax forests, may begin to appear during the intermediate tree stage. In general, however, the habitat appears to be rather impoverished with regard to its animal life, especially during its early years.

FIELD WORK IN THE INTERMEDIATE TREE COMMUNITY

Quadrats of five to ten meters square (preferably ten) are suitable for this community. Collections of invertebrates and observations of vertebrates should follow the same pattern as in earlier field trips.

What plant groups are most abundant here? What species are dominant? What ecological factors have entered into the picture to modify plant life? Is there any change in the upper few inches of soil? Are the tree seedlings of the same species as the full-grown trees? Are there evidences of sprouting among the older trees? (Multiple stems are a good indication that trees have developed as sprouts from stumps.) How do you explain this phenomenon? What will be the result?

How does the invertebrate population compare with that in earlier stages? Can you construct a food chain on the basis of the animals observed? Look under fallen logs for runways, nests or other signs of small mammal activity. Did your trapline take any species not taken in earlier stages? If so, can you suggest what ecological changes might have affected their distribution? Estimate the length of time which has been occupied by the process of succession up to this stage. Have you any definite evidence for this estimate?

REFERENCES

Bramble, William C. and Roger H. Ashley. 1955. Natural revegetation of spoil banks in central Pennsylvania. Ecology 36:417-423.

Brewer, Richard and Edward D. Triner. 1956. Vegetational features of some strip-mined land in Perry County, Illinois. Illinois Acad. Sci. 48:73-84.

Buell, Murray F. and F. H. Bormann. 1955. Deciduous forests of Ponemah Point, Red Lake Indian Reservation, Minnesota. Ecology 36:646-658.

Godman, Richard M. and Laurits W. Krefting. 1960. Factors important to yellow birch establishment in upper Michigan. Ecology 41: 18-28.

Martin, N. D. 1959. An analysis of forest succession in Algonquin Park, Ontario. Ecol. Monogr. 29:187-218.

Tevis, Lloyd, Jr. 1956. Responses of small mammal populations to logging of Douglas fir. J. Mammal. 37:189-196.

See ALSO references on pp. 77-78.

For identification manuals, see pp. 218-226.

The Climax Forest

In various parts of the North American continent, many different types of climax forests occur. Three major types are generally listed by ecologists: coniferous forests, deciduous forests, and tropical rain forest. Within each of these major groupings, however, there are a great many variations. Over much of the forested region of Canada, a variety of forests dominated by conifers will be found. In the northeastern and central United States, a deciduous forest occurs, with its dominant trees varying from beech and sugar maple around the Great Lakes to oaks near the edges of the prairie. In the southeast a forest of conifers mixed with hardwoods covers large areas, while in much of Mexico and Central America tropical rain forests exist. The Rocky Mountain region shows great altitudinal variation in climax type, with many species of both hardwoods and conifers represented at various elevations. Finally, the Pacific Coastal Region includes several different climax communities.

It is evident, then, that it is not possible to generalize with regard to the actual species which you will find in a climax forest. We may, however, make some generalizations about the types of plants and animals which may be expected in such vegetational types. In coniferous forests which are dense in character, undergrowth is often sparse. In more open forests, many shrubs such as species of dogwood, viburnum, heaths, maples, yew, juniper and others will be found. Similarly, the herbaceous growth in more open forests is abundant. Spring flowering species such as violets, triliums and orchids are interspersed with a variety of ferns. The characteristic layer of humus (decayed organic matter) is built up by fallen leaves, rotting vegetation, dead roots and branches. This organic matter provides fertility and also retains large quantities of water, thus supplying large numbers of plants with the life-giving fluid even in relatively dry seasons. Many forests contain extensive swamps, where standing water may persist throughout much of the year, gradually giving up water during the summer to the air and to the surrounding forest.

The climax forest is typically rich in animal life, although much of it is not obvious. Insects are abundant everywhere, from the ground underfoot to the tops of the trees; inside fungi and wood; and in many other situations. Spiders and other predators feed on these insects. The invertebrate fauna of the forest floor is rich and varied, including such obvious species as centipedes, sowbugs, and earthworms, and many less obvious ones.

All classes of terrestrial vertebrates occur in the forest. Salamanders are often common under dead logs and stones, especially in moist locations. Some species are found in standing dead trees, while others spend most of their time underground. The tree-hole habitat is sometimes occupied by tree frogs, and wood frogs may be seen on the ground. Although the reptile fauna of a woodland may not be as rich as that of a prairie or desert location, a few species of snakes will be found in most forests. These may be predators upon the frogs and salamanders, as well as upon insects and worms. One species of turtle, the wood turtle, is typically found in forests within its range, while such lizards as skinks and iguanids may live in this habitat.

The variety of habitat niches in the forest provides for a large number and immense variety of birds. Woodpeckers drill in dead trees, jays and flycatchers, warblers, vireos and grosbeaks forage through the trees or catch flying insects. Thrushes and some warblers occupy the ground habitat, while predatory birds, hawks and owls, play their part in the total ecology of the forest.

Forest mammals range in size from the tiny pigmy shrew to the giant moose and grizzly bear. The forest floor is the home of rodents and insectivores, often in considerable variety. In the trees, squirrels make their homes and tree bats may spend the day hanging from branches or inside hollow trees. These smaller mammals are fed upon by weasels, martens, raccoons and other predators. The larger herbivores, the deer, rabbits, wild pigs and such, are prey for bobcats, panthers, wolves and other large predators. Many of the larger organisms have been extirpated from much of their former range, but smaller ones, up to the size of the bobcat and the deer, persist even in well-populated areas.

FIELD WORK IN THE FOREST COMMUNITY

In a forest, a transect (see p. 71) is often used instead of quadrats. With a compass, establish a straight line through the forest, and tabulate all trees and large shrubs within a given distance of this base line. Quadrats may be set up for detailed study of the smaller vegetation. Animals and animal signs should be observed carefully. Look for dead standing trees and examine them for signs of nests, holes, roosting bats or other life. Turn over logs and stones and observe the life underneath them. If it is the nesting season, look for bird nests at all levels of the vegetation. Spread a large canvas under a tree small enough to be shaken vigorously, and see what species of insects can be shaken out onto the canvas.

What orders and families of insects are most abundant? Did you secure an adequate sample on which to base a conclusion? Are the insect groups different from those represented in earlier stages of succession? Arrange the tree species in order of their numerical abundance. Is their importance in the community in the same order as their abundance? Are the same shrubs here that you found in the shrub stage of succession? Would you say that animal life is more abundant here than in earlier stages of succession? Is this true in numbers, in mass, in both or neither?

REFERENCES

Buell, Murray F. and William A. Niering. 1957. Fir-spruce-birch forest in northern Minnesota. Ecology 38:602-610.

Daubenmire, R. F. 1943. Vegetational zonation in the Rocky Mountains. Botan. Rev. 9:325-393.

____________1952. Forest vegetation of northern Idaho and adjacent Washington, and its bearings on concepts of vegetation classification. Ecol. Monogr. 22:301-330.

Davis, J. H., Jr. 1943. The natural features of southern Florida, especially the vegetation and the Everglades. Florida Geol. Survey Bull. 25:5-311.

Kenoyer, L. A. 1929. General and successional ecology of the lower tropical rain forest at Barro Colorado Island, Panama. Ecology 10:201-222.

Quarterman, Elsie, and Catherine Keever. 1962. Southern mixed hardwood forest: climax in the southeastern coastal plain, USA. Ecol. Monogr. 32:167-185.

Shelford, V. E. and S. Olson. 1935. Sere climax and influent animals with special reference to the transcontinental coniferous forest of North America. Ecology 16:375-402.

Whittaker, Robert H. 1956. Vegetation of the Great Smoky Mountains. Ecol. Monogr. 26:1-80.

____________1960. Vegetation of the Siskiyou Mountains, Oregon and California. Ecol. Monogr. 30:279-338.

See ALSO references on pp. 77-78.
For identification manuals, see pp. 218-226.

Grassland Climax

In areas where rainfall is limited, especially during the hot summer months, a climax community develops in which grasses are the dominant forms. A great part of the mid-west from the Mississippi to the Rocky Mountains originally had a grassland climax, or prairie. Since prairie is usually excellent farmland, it is now difficult to find any of the original, undisturbed prairie. At present, the right of way along railroads often provides the best example of this climax type.

The kinds of grasses that may predominate in any one locality vary. The eastern part of the prairie is said to be a tall grass, the western part short grass prairie. Furthermore, wet areas will possess a distinctive community, quite different from that found in drier areas. Tall grass prairies are often composed of such grasses as cordgrass, Spartina pectinata; Indian grass, Sorghastrum nutans; tall oat grass, Arrhenatherum elatius; manna grasses, Glyceria sp.; panic grasses, Panicum sp.; bluestems, Andropogon sp.; porcupine grass, Stipa spartea; and dropseed, Sporobolus sp. Some of these species may grow to heights of 6 to 8 feet. Short grass prairies may include grasses like buffalo grass, Buchloe dactyloides, grama grass, Bouteloua sp., and wheat grass, Agropyron sp.

Animals were once fairly numerous on the prairie. In fact, some ecologists believe the buffalo was the dominant species of all organisms. In addition, badgers, prairie dogs, coyotes, antelope, and ground squirrels were also to be found there. Naturally, remnants of prairie along railroad tracks cannot support many of these forms. Still to be found are prairie deer mice, meadow mice, ground squirrels, and most of the amphibians, reptiles, birds, and invertebrates of former times. Many birds inhabit the region, some of them being characteristic of this climax. These include the upland plover, prairie horned lark, Bell's vireo, dickcissel, chestnut-collared longspur and long-billed curlew.

The vegetation of this climax undoubtedly affects the distribution of the animals. Some birds apparently require trees in which to build their nests, while others are able to nest on the ground in grassy areas. The summer heat unrelieved by the shade of trees, and the windy nature of the plains may also be restrictive to certain animals. Some species requiring trees may still be able to exist in this region as inhabitants of the wooded river-bottoms. However, this region is mostly flat, and without many permanent streams.

Lack of such sources of water also has its direct limiting effects on animal distribution. Agricultural practices also tend to dry out the land, and drought may reduce populations drastically. For these reasons, species may be represented by only a few individuals or may be absent from large areas of their possible range during certain years.

FIELD WORK IN THE GRASSLAND CLIMAX

Establish one or more quadrats one meter square, and count all the plants within the squares. Record this information on a quadrat data sheet (page 103). Note the approximate heights of each kind of grass found in the quadrat. Are the plants crowded together forming a compact sod, or is there open ground between clumps of grass Is there a great variation in plant types between wet and dry areas? Are there any woody plants present?

Traps set out 24 hours previously may be examined for the day's catch. Are there any signs of mammals, such as scats, rabbit trails or forms? If any damp areas are in the vicinity, look for crayfish chimneys and presence of amphibians. Sweep the vegetation for invertebrates and record the kinds and relative numbers caught.

What is the nature of the soil - dark or light brown, loam, sand or clay? Soil maps of the area may give you the exact type. Determine the pH of the soil. Is the soil in general well-drained or not? What is the depth of the A horizon (dark soil having large amounts of decayed organic matter)?

From the observations made, how can you determine whether or not this is a true climax, or only a grass stage in a succession toward forest? Is the prairie you have studied a tall grass or short grass prairie? Why are the A horizons of prairie soils so deep and rich? What is one possible reason why woody plants are not typical prairie forms? It is sometimes said that buffalo and fires set by Indians helped to maintain the prairie, at least on its Eastern border. Could you give any reason for agreeing or disagreeing with this statement?

REFERENCES

Ayyad, M. A. G. and R. L. Dix. 1964. An analysis of a vegetation-microenvironmental complex on prairie slopes in Saskatchewan. Ecol. Monogr. 34:421-442.

Burcham, L. T. 1957. Calfornia range land: an historico-ecological study of the range resource of California. Dept. Nat. Res., Sacramento. 261 pp.

Carpenter, J. Richard. 1940. The grassland biome. Ecol. Monogr. 10:617-684.

Coupland, Robert T. 1958. The effects of fluctuations in weather upon the grasslands of the Great Plains. Botan. Rev. 24:273-317.

________________ 1961. A reconsideration of grassland classification in the northern Great Plains of North America. J. Ecology 49:135-167.

Crockett, Jerry. 1964. Influence of soils and parent materials on grasslands of the Wichita Mountains Wildlife Refuge, Oklahoma. Ecology 45:326-335.

Curtis, J. T. 1955. A prairie continuum in Wisconsin. Ecology 36: 558-566.

Ellison, Lincoln. 1960. Influence of grazing on plant succession of rangelands. Botan. Rev. 26:1-78.

Fichter, E. 1954. An ecological study of invertebrates of grassland and deciduous shrub savanna in eastern Nebraska. Amer. Midland Naturalist 51:321-339.

Hayes, W. P. 1927. Prairie insects. Ecology 8:238-250.

Malin, J. C. 1956. The grassland of North America: its occupance and the challenge of continuous reappraisals. In Man's role in changing the face of the earth, ed. by W. L. Thomas. Univ. of Chicago Press. pp. 350-366.
Mentzer, Loren. W. 1951. Studies on plant succession in true prairie Ecol. Monogr. 21:255-267.
Rice, Elroy. 1964. Inhibition of nitrogen-fixing and nitrifying bacteria by seed plants. Ecology 45:824-837.
Sprague, Howard B. (ed.) 1959. Grasslands. Am. Assoc. for the Advanc. of Sci., Washington, D. C.
Tester, John R. and William H. Marshall. 1961. A study of certain plant and animal interrelations on a native prairie in northwestern Minnesota. Univ. of Minn. Press, Minneapolis. 51 pp.
Weaver, J. E. 1954. North American prairie. Johnson Publ. Co., Lincoln, Nebr.
______________ and F. W. Albertson. 1956. Grasslands of the Great Plains. Johnson Publ. Co., Lincoln, Nebr.
See ALSO references on pp. 77-78.
For identification manuals, see pp. 218-226.

Desert Climax

Where climates are warm and drought is the normal situation, a community known as a desert climax becomes established. As with forest and grassland climaxes, there are different kinds of desert climaxes, depending on regional variations of rainfall, drainage, temperature, and other factors. The American Desert occupies most of the area between the Rocky Mountains and the Sierra Nevada Mountains, and at the lower elevations in this region, examples of varying kinds of desert climaxes may be found.

Plants of the desert must in general be able to survive on limited amounts of water, often depending only on a short rainy season for the production of flowers and seeds. Typical desert plants include sagebrush, Artemia sp.; grease wood, Sarcobatus vermiculatus; creosote bush, Larrea divaricata; mesquite, Prosopis chilensis; shadscale, Atriplex sp.; iodine bush, Allenrolfea occidentalis; and Joshua trees, Yucca brevifolia. Common cacti include saguaros, chollas, prickly pears, and barrel cacti. Yuccas, ocotillo, salt blite, burro weed, palo verde and the century plant are other characteristic plants of desert climax communities.

Animals of the desert are more abundant than is ordinarily suspected. Many of them are nocturnal, and thus escape normal observation. Birds, lizards, and a few insects can be seen during the daytime. At night, however, a greater variety of animals can be observed. Kangaroo rats, kit foxes, ringtail cats, jackrabbits, mice, rattlesnakes, scorpions and other invertebrates are among the nocturnal animals that are able to survive the rigors of this habitat.

FIELD WORK IN THE DESERT CLIMAX

It is frequently wise to plan field trips to the desert so that the class arrives at the study site in the late afternoon. This allows time for examination of the area for plants and diurnal animals, and also to set out traps for nocturnal mammals which are more abundant. The trip may conclude late at night or early in the morning. Due precautions should be taken against poisonous animals.

Record the temperature and relative humidity at half hour intervals. Note the hour of sundown and darkness.

Select areas for large quadrats (10 meters) or use a transect to sample the plants of this climax, and record the numbers of each species found. Which species appear to be dominant? Are the plants evenly distributed, or more or less gathered in clumps? What is the explanation for this? Are any of the plants indicators of special physical conditions?

Examine the area for animal signs. Rat houses may be found in the center of brush clumps. Lizards may also be flushed out of such areas. Are any insects to be found on the vegetation? Birds may be seen especially at sundown and in the early evening hours. If the wind has not been active, invertebrate and vertebrate tracks may be seen in sandy areas. Bats may be seen at twilight, especially near the bases of hills or near stream beds.

A good catch of small mammals with snapback traps in a deciduous forest is about 10%, and catches are frequently less than this. On the basis of the animals caught on this trip, would you say that this desert climax supports a higher or lower small mammal population than the deciduous forest? How do the plants survive under the drought conditions of the desert?

REFERENCES

Benson, Lyman and Robert Darrow. 1954. Trees and shrubs of the southwestern deserts. Univ. of New Mexico Press, Albuquerque.

Buffington, Lee C., and Carlton H. Herbel. 1965. Vegetational changes on a semi-desert grassland range. Ecol. Monogr. 35: 139-164.

Buxton, P. A. 1923. Animal life in deserts. St. Martin's Press, N. Y.

Cloudsley-Thompson, J. L. 1954. Biology of deserts. Hafner Publishing Co., N. Y.

Fautin, Reed. 1946. Biotic communities of the northern shrub biome in western Utah. Ecol. Monogr. 16:251-310.
Jaeger, Edmund C. 1957. The North American deserts. Stanford Univ. Press, Stanford.
Kirmiz, J. P. 1962. Adaptation to desert environment. Butterworth, Inc., Washington, D. C.
Shields, Lora M. and Linton Gardner (eds.) 1961. Bioecology of the arid and semiarid lands of the southwest. New Mexico Highlands Univ., Las Vegas. Bull. 212.
Shreve, Forrest. 1942. The desert vegetation of North America. Botan. Rev. 8:195-246.
__________ and Ira L. Wiggins. 1964. Vegetation and flora of the sonoran desert. 2 vols. Stanford Univ. Press, Stanford.
See ALSO references on pp. 77-78.
For identification manuals, see pp. 218-226.

Microsuccession in Rotten Logs

The principle of succession may be demonstrated in microhabitats within a community, such as plant galls, fecal droppings of large mammals, and rotting logs. In each case, the microhabitat undergoes physical and chemical changes which accomplish its destruction, and its remains become a part of the soil of the community. The physical and chemical changes are brought about by biotic forces primarily. Bacteria, fungi, invertebrates, and vertebrates all aid in changing the original microhabitat.

In the case of the rotting log microsuccession, the changes may begin while the dead tree is still standing. At that time certain insects and other invertebrates may inhabit the bark and outer wood. Birds and mammals may use hollow parts for nests.

The tree finally falls to the ground after being weakened by the boring insects and the rotting effects of bacteria and fungi. The bark is probably already off, and thus the first inhabitants of the tree will now be replaced by new ones. The wood will be further riddled with boring, and more fungi may gain entrance to the inner parts of the wood.

The plants and animals change the wood both physically and chemically, until once again, a new kind of community is found in the log. At this stage, the inside of the log may be "punky", while the outer shell remains firm. In such a condition, logs often serve as homes for small mammals such as the common shrew and white footed mice, as well as lizards and salamanders.

Eventually, even this outer shell disintegrates under the attack of organisms, and the log, now nearly a part of the forest

floor, plays host to still a different set of organisms. For example, many of the smaller snakes like the ring-necked and worm snake find this habitat suitable to their needs. These will finally disappear to be replaced by the characteristic inhabitants of the forest floor, as the log is completely decayed.

FIELD WORK IN ROTTEN LOG MICROSUCCESSION

In any one log, the succession from the dead tree stage to the stage where the log has become part of the forest floor takes several years. Observation of this succession may be witnessed in a few hours by examining several logs of the same species in the same community, at different stages of decay. The instructor will indicate trees representing stages in the microsuccession, and the class may then make various observations on these stages.

Stage I Standing dead tree.

Is there bark on the tree? If so, it is easily removed or not? Is the wood hard and dry? What invertebrates can you find under the bark, or in the wood? Are wood borers present? Are any vertebrates, such as squirrels or birds, nesting in the tree? Record all species of animals found.

Stage II Newly fallen tree.

Is there any bark on the tree at this stage? Is the wood firm or soft, wet or dry? What invertebrates are found in it or in the wood? Are any borers present? (An axe may be necessary here to observe deep tunnels in the wood.)

List all species and their relative abundance.

Stage III Log rotting inside, but hard on outside.

Lift off the outer shell, and closely examine the contents. Break apart the shell, being careful to note all invertebrates. Rake through the punky part of the log, being especially watchful for lizards, snakes, and salamanders, as well as invertebrates. Record all species and their numbers. Are there any mammal runways in or under the log?

Stage IV Completely rotten log.

Rake through the rotten wood as in the stage three log, noting all species observed and their quantities. Is the wood more moist or less so than in the previous stage?

In which stage were the most kinds of animals found? In which stage were the most animals of all species found? What is the most striking physical difference between the first and the last stage of the succession? How can you describe the chemical difference between the first and last stages?

REFERENCES

Allee, W. C., A. E. Emerson, O. Park, T. Park, and K. P. Schmidt. 1949. Principles of animal ecology. W. B. Saunders Co., Philadelphia.

Blackman, M. W. and S. S. Stage. 1924. On succession of insects living in the bark and wood of dying, dead, and decaying hickory. N. Y. S. Coll. of Forestry. Tech. Bull. 17:1-269.

Holmquist, A. M. 1926. Studies in arthropod hibernation. Ann. Entom. Soc. of Amer. 19:395-428.

Mohr, C. O. 1943. Cattle droppings as ecological units. Ecol. Monogr. 13:275-298.

Park, Orlando, 1931. Studies in the ecology of forest Coleoptera II. The relation of certain Coleoptera to plants for food and shelter, especially those species associated with fungi in the Chicago Area. Ecology 12:188-207.

Savely, H. E. 1939. Ecological relations of certain animals in dead pine and oak logs. Ecol. Monogr. 9:321-385.

Shelford, Victor E. 1913. Animal communities in temperate America. Bull. Geog. Soc. Chicago 5:1-368. Reprinted 1937 Univ. of Chicago Press.

For identification manuals, see pp. 218-226.

ROTTEN LOG MICROSUCCESSION
DATA SHEET
(page 1)

STUDENT: ____________________ DATE: ____________

Location ____

Soil type: __________ Soil temp.: ________ Air temp.: ______

Rel. humidity: ________ Vegetational type: ______________

STAGE I		STAGE II	
Condition of log:		Condition of log:	
No.	Species	No.	Species

ROTTEN LOG MICROSUCCESSION DATA SHEET (page 2)

STUDENT:____________________________ DATE:__________________

STAGE III		STAGE IV	
Condition of log:		Condition of log:	
No.	Species	No.	Species

QUADRAT DATA SHEET

STUDENT:______________________ DATE:______________

Locality:______________________ Vegetational type:______________

Soil type:____________ Depth of A horizon:__________ Soil pH.__________

Soil temp:________ Air temp:________ Rel. humidity:__________

Quadrat No.________

Species	No.	% Area	Growth type*	Importance**

* Growth type: A = annual herb; P = perennial herb; S = shrub; T = tree.

** Importance: D = dominant; C = common; U = uncommon; R = rare.

QUADRAT DATA SHEET

STUDENT:______________________________ DATE:__________________

Locality:_______________________________ Vegetational type:__________________

Soil type:_______________ Depth of A horizon: ___________ Soil pH.____________

Soil temp:_________ Air temp:__________Rel. humidity:____________

Quadrat No._________

Species	No.	% Area	Growth type*	Importance**

* Growth type: A = annual herb; P = perennial herb; S = shrub; T = tree.

** Importance: D = dominant; C = common; U = uncommon; R = rare.

QUADRAT DATA SHEET

STUDENT:______________________ DATE:______________

Locality:______________________ Vegetational type:______________

Soil type:____________ Depth of A horizon:__________ Soil pH.__________

Soil temp:________ Air temp:________ Rel. humidity:__________

Quadrat No.________

Species	No.	% Area	Growth type*	Importance**

* Growth type: A = annual herb; P = perennial herb; S = shrub; T = tree.

** Importance: D = dominant; C = common; U = uncommon; R = rare.

QUADRAT DATA SHEET

STUDENT:______________________________ DATE:________________

Locality:______________________________ Vegetational type:________________

Soil type:______________ Depth of A horizon:__________ Soil pH.____________

Soil temp:________ Air temp:__________ Rel. humidity:____________

Quadrat No.________

Species	No.	% Area	Growth type*	Importance**

* Growth type: A = annual herb; P = perennial herb; S = shrub; T = tree.

** Importance: D = dominant; C = common; U = uncommon; R = rare.

TRANSECT DATA SHEET

STUDENT:____________________ DATE:____________

Locality:____________________ Vegetational type:____________

Soil type:__________ Depth of A horizon:________ Soil pH:________

Soil temp.:______ Air temp.:______ Rel. Humidity:________

Width of transect:__________ Length of transect: ________

Species	No.	% Area	Growth type *	Importance **

* Growth type: A = annual herb; P = perennial herb; S = shrub; T = tree.

** Importance: D = dominant; C = common; U = uncommon; R = rare.

TRANSECT DATA SHEET

STUDENT: ______________________ DATE: ______________

Locality: ______________________ Vegetational type: ______________

Soil type: ____________ Depth of A horizon: ____________ Soil pH: ____________

Soil temp.: ________ Air temp.: ________ Rel. Humidity: ____________

Width of transect: ______________ Length of transect: ____________

Species	No.	% Area	Growth type *	Importance **

* Growth type: A = annual herb; P = perennial herb; S = shrub; T = tree.

** Importance: D = dominant; C = common; U = uncommon; R = rare.

TRANSECT DATA SHEET

STUDENT:______________________________ DATE:________________

Locality:______________________________ Vegetational type:________________

Soil type:______________ Depth of A horizon:____________ Soil pH:____________

Soil temp.:________ Air temp.:________ Rel. Humidity:____________

Width of transect:______________ Length of transect: __________

Species	No.	% Area	Growth type *	Importance **

* Growth type: A = annual herb; P = perennial herb; S = shrub; T = tree.

** Importance: D = dominant; C = common; U = uncommon; R = rare.

TRANSECT DATA SHEET

STUDENT:______________________ DATE:______________

Locality:______________________ Vegetational type:______________

Soil type:__________ Depth of A horizon:__________ Soil pH:__________

Soil temp.:________ Air temp.:________ Rel. Humidity:__________

Width of transect:__________ Length of transect: __________

Species	No.	% Area	Growth type *	Importance **

* Growth type: A = annual herb; P = perennial herb; S = shrub; T = tree.

** Importance: D = dominant; C = common; U = uncommon; R = rare.

Section IV
Aquatic Communities and Succession

Introduction

The same process of succession that may be seen in certain terrestrial situations also occurs in aquatic environments. Deep lakes evolve into shallow lakes, then into ponds, bogs or swamps. Eventually, the area that once was a deep lake may possess a terrestrial climax community. Similarly, streams change from swift to slow-flowing streams, and perhaps eventually to dry land. As these aquatic environments evolve, the communities within them change, and help to bring about future developments.

To observe the changes that may occur in aquatic situations, one merely has to select different lakes, ponds, or streams as representative stages of these successions. A study of such communities will reveal the differences in the species composition of each, and how the physical conditions of the habitat affect this composition. It will also demonstrate how these successions occur, and the extent to which physical and biological forces are involved in their accomplishment. Although aquatic succession and terrestrial succession have certain basic similarities, it should be noted at the outset that the relative importance of the various ecological factors is quite different in the two situations.

REFERENCES

Benton, Allen H. and William E. Werner, Jr. 1965. Field biology and ecology. McGraw-Hill Book Co., N. Y.

Coker, Robert E. 1954. Streams, lakes, ponds. Univ. of North Carolina Press, Chapel Hill, N. C.

Edmonson, W. T. (ed.) 1959. Ward and Whipple's fresh water biology. Second ed. John Wiley and Sons, N. Y.

Frey, David G. (ed.) 1963. Limnology in North America. Univ. of Wisconsin Press, Madison.

Hutchinson, G. Evelyn. 1957. A treatise on limnology, Vol. I: geography, physics, and chemistry. John Wiley and Sons, N. Y.

Langlois, Thomas H. 1954. The western end of Lake Erie and its ecology. J. W. Edwards, Ann Arbor, Mich.
Macan, T. T. 1963. Freshwater ecology. John Wiley and Sons, N. Y.
__________, and E. B. Worthington. 1951. Life in lakes and rivers. Collins, London.
Morgan, Ann H. 1930. Field book of ponds and streams. G. P. Putnam's Sons, N. Y.
Needham, J. G. and J. T. Lloyd. 1937. The life of inland waters. Comstock Publ. Co., Ithaca.
__________ and Paul R. Needham. 1962. A guide to the study of freshwater biology. Holden-Day Inc., San Francisco, Calif.
Nikolsky, G. V. 1962. Ecology of fishes. Academic Press, N. Y.
Pennak, R. W. 1953. Fresh water invertebrates of the United States. Ronald Press, N. Y.
Reid, George K. 1961. Ecology of inland waters and estuaries. Reinhold Publ. Co., N. Y.
Ruttner, F. 1963. Fundamentals of limnology. University of Toronto Press, Toronto, Ontario.
Welch, Paul S. 1952. Limnology. McGraw-Hill Book Co., N. Y.
For films and film strips on aquatic communities, see p. 229.

Lake Community

The first stage in the succession of standing bodies of water is the deep lake community. Such lakes usually stratify thermally during mid-summer. At that time, there will be a region or layer of water in which the temperature decreases at least 1° C. for each meter of increasing depth. This region is called the thermocline, and is located well below the surface. Above this region is the epilimnion, and below the hypolimnion. The open water is called the limnetic zone, down to the depth at which light no longer penetrates in sufficient quantity for plants to be able to manufacture as much oxygen as they utilize in respiration (compensation point). The region below this is called the profundal zone. The organisms living on the bottom constitute the benthos, (e.g., dragon fly larvae, midge larvae), while those swimming through the water are called the nekton (e.g., fishes, turtles), and those living at the surface the neuston (e.g., gyrinid beetles, water skippers). Organisms in the water too small to be able to determine the direction of their own movements are termed plankton (e.g., protozoa, algae).

If a lake is very deep, the waters of the hypolimnion usually possess oxygen at any season. Such lakes do not produce a large amount of organisms for their volume, and are called oligotrophic (little producing) lakes. They frequently have only small shore areas in which aquatic vegetation may grow. However, their profundal and benthic fauna are abundant. Such lakes fill in or drain, resulting in shallower lakes. These shallower lakes have a rich

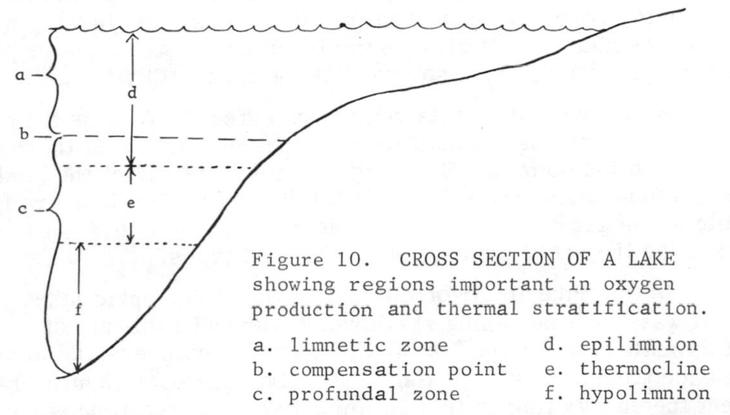

Figure 10. CROSS SECTION OF A LAKE showing regions important in oxygen production and thermal stratification.

a. limnetic zone
b. compensation point
c. profundal zone
d. epilimnion
e. thermocline
f. hypolimnion

biota, and thus are called eutrophic (good producing). Their hypolimnion may become depleted of oxygen before the thermocline is destroyed at the end of summer. For that reason the benthos in the profundal zone may be impoverished, although the other areas of the lake are highly productive. Frequently such lakes also have extensive areas of rooted aquatic plants. Warm water fish may be abundant.

As in the study of other communities, any one field trip will provide only a fragmentary picture of life in it, for seasonal and annual changes in the biota may be considerable. This fact should be kept in mind in drawing conclusions about the habitat studied.

FIELD WORK IN THE LAKE COMMUNITY

Before the water is disturbed by collecting, obtain samples of water at depths indicated by the instructor, and determine oxygen and carbon dioxide concentrations, and pH. Obtain temperature readings at these same depths. If hydrologic maps are available, determine the maximum depth of the lake.

After the water samples are obtained, collect plankton samples, using a plankton net. The plankton should be studied later in the laboratory for kinds and numbers of plants and animals present.

Ten-foot seines may be operated along the shore to sample the vertebrate fauna of the littoral (shore-line) zone. Dip nets may be used to collect invertebrates. The benthos of the shore may be collected and observed by using seives to strain organisms from the shore bottom. White pans are useful to help observe small invertebrates. Note the kinds present and their relative abundance. Search for frogs and salamanders along the shore, turning over all stones and logs. Collect samples of the rooted aquatic plants for identification, and also examine them carefully for animals.

How far out does this vegetation extend? What is the depth of the water at this outward limit of the vegetaion? Is there organic matter on the bottom? How might bottom type affect the kinds of plants found along the shore? What physical conditions are favorable to the growth of rooted aquatic plants, other than bottom type? Examine the shore plants, noting unusual types.

Is this lake an oligotrophic lake or an eutrophic lake? In what ways is it becoming shallower? How will this affect the kinds of animals found in the lake? Compare the numbers and biomass of each trophic level. (See p. 121.) Do your data accurately represent the energy flow in this community? (See references on p. 115-116 and p. 138.)

REFERENCES

Forbes, S. A. 1925. The lake as a microcosm. Illinois Nat. Hist. Survey Bull. 15:537-550.

Krecker, Frederick H. 1931. Vertical oscillations or seiches in lakes as a factor in the aquatic environment. Ecology 12: 156-163.

Langlois, Thomas H. 1954. The western end of Lake Erie and its ecology. J. W. Edwards, Publisher, Ann Arbor, Mich.

Mandossian, Adrienne, and Robert P. McIntosh. 1960. Vegetation zonation on the shore of a small lake. Am. Midland Naturalist 64:301-308.

Megard, Robert O. 1964. Biostratigraphic history of Dead Man Lake, Chuska Mountains, New Mexico. Ecology 45:529-546.

Moore, Walter G. 1950. Limnological studies of Louisiana lakes. I. Lake Providence. Ecology 31:86-99.

Mortimer, C. H. 1941. The exchange of dissolved substances between mud and water in lakes. J. Ecology 29:280-329.

Moulton, F. R. (ed.) 1939. Problems of lake biology. Am. Assoc. Adv. Sci. Pub. No. 10. Science Press, Lancaster, Pa.

Oosting, Henry J. 1933. Physical-chemical variables in a Minnesota lake. Ecol. Monogr. 3:493-533.

Pennak, R. W. 1949. Annual limnological cycles in some Colorado reservoir lakes. Ecol. Monogr. 19:233-267.

Tucker, Allan. 1957. The relation of phytoplankton periodicity to the nature of the physico-chemical environment with special reference to phosphorus. Am. Midland Naturalist 57:300-370.

Vallentyne, J. R. 1957. The principles of modern limnology. Am. Scientist 45:218-244.

Wilson, L. R. 1935. Lake development and plant succession in Vilas County, Wisconsin. I. Medium hard water lakes. Ecol. Monogr. 5:207-247.
See ALSO references on pp. 115-116.
For identification manuals, see pps. 218-226.

Pond Community

As lakes fill in, they eventually reach a stage at which the water is so shallow that any thermal stratification is quickly destroyed by the wind, and thermoclines never form. The rooted aquatic vegetation becomes more conspicuous, occupying a greater per cent of the bottom area of the lake. It may eventually extend entirely across the lake; under these circumstances, the lake is usually called a pond. It will still be very productive, and may especially harbor many species of amphibians, reptiles, and invertebrates. The benthos will contain abundant organisms in all regions.

Due to the shallow nature of the pond, the temperature will be more like that of the air, becoming warm in the summer, and freezing over quickly in the winter (in the colder climates). The many animals and decaying organisms present may use up all the oxygen under the ice, resulting in suffocation of the occupants.

The abundant vegetation will hasten the final filling of the pond. As it approaches extinction, it may dry up during the dry months, but have enough water in it to support organisms during the wet season. In such instances, the organisms must be able to form spores, migrate, or withstand drought, and this creates an unusual community, known as the temporary pond community.

The shores of ponds will reveal the future development of the pond, for as the pond fills in the shoreline approaches the spot that used to be the center of the open water. From the open water to the shoreline, the following plant zones or communities may be observed: submerged aquatics, floating aquatics, emergents. From the shoreline to the climax community around the pond, the following zones may be seen: rank herbaceous vegetation, shrub, temporary tree, and climax forest. As succession proceeds, each community will replace the one next to it toward the center of the pond. The final stage in the standing water succession will be completed when the climax community covers the area once occupied by water.

FIELD WORK IN THE POND COMMUNITY

Obtain water samples from locations as directed by the instructor, and from them determine the oxygen and carbon dioxide concentrations, and pH. Also note the temperature readings from these locations. Determine the depth of the water.

Collect plankton samples, to be examined for composition later. Numbers and kinds of species present should be noted. Dip nets may be used to collect large invertebrates, and seines for fish, amphibians and turtles. Wire net scoops are useful for straining organisms from mud bottoms. Note kinds and abundance of all animals found.

Make careful observations on the plant life present, from the submerged aquatics in toward shore, including the floating aquatics and emergent aquatics. Also observe the zones of communities on shore, from shoreline back to the climax vegetation (if any) around the pond. What are the dominant plants of each zone? In what ways are the physical conditions of each zone different from the preceding zones? Is this pond relatively productive? What factors may affect its productivity? How will its productivity affect the rate of succession of the pond to dry land? List the kinds and numbers of organisms in each trophic level (see p. 121). Does this represent the true picture of energy flow through the community?

REFERENCES

Conover, Robert J. 1961. A study of Charlestown and Green Hill Ponds, Rhode Island. Ecology 42:119-140. (Brackish-water ponds)

Dickinson, J. C., Jr. 1949. An ecological reconaissance of the biota of some ponds and ditches in northern Florida. Quart. J. Florida Acad. Sci. 11:1-28.

Dineen, C. T. 1953. An ecological study of a Minnesota pond. Am. Midland Naturalist 50:349-376.

Judd, W. W. 1960. A study of the population of insects emerging as adults from South Walker Pond at London, Ontario. Am. Midland Naturalist 63:194-210.

Paloumpis, A. A. 1957. The effects of drought on the fish and bottom organisms of two small oxbow ponds. Trans. Illinois Acad. Sci. 50:60-64.

Shelford, V. E. 1911. Ecological succession: pond fishes. Biol. Bull. 21:121-151.

See ALSO references on pp. 115-116.
For identification manuals, see pp. 218-226.

PYRAMID OF TROPHIC LEVELS IN FRESHWATER COMMUNITIES

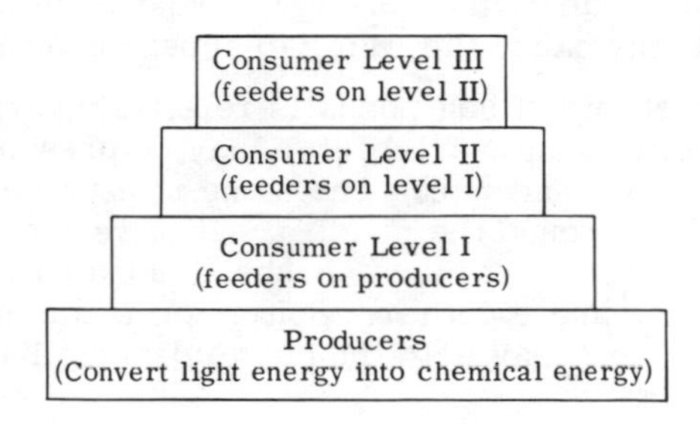

Examples of organisms belonging to the various trophic levels:

<u>Producers</u>

- Photosynthetic bacteria
- Floating algae
- Attached algae
- Bryophytes
- Submerged, floating, and emergent vascular plants.

<u>Scavengers</u>

Feeders on dead organic matter, often in particulate form, from all trophic levels

- Bacteria
- Protozoa
- Planaria
- Nematodes
- Rotifers
- Bryozoans
- Annelids
- Cladocera
- Copepods
- Ostracods
- Amphipods
- Black fly larva
- Clams

<u>Consumer Level I</u>

Feeders on phytoplankton

- Protozoa
- Sponges
- Rotifers
- Tardigrades
- Bryozoans
- Cladocerans
- Copepods
- Haliplid beetles
- Mosquito larva
- Clams
- Snails
- Some fish (e.g., gizzard shad)

Feeders on larger plants-Bryophytes and vascular plants

- Nematodes
- Crayfish
- Mayfly larva
- Snails
- Ducks
- Muskrats, beavers

<u>Consumer Level II</u>

Feeders on zooplankton

- Planaria
- Rotifers
- Nematodes
- Cladocerans
- Copepods
- Blackfly larva
- Mosquito larva
- Juvenile fish

Feeders on larger invertebrates

- Dytiscid beetles
- Odonate nymphs
- Water scorpions
- Dobson flies
- Water bugs
- Small fish
- Frogs, salamanders
- Turtles
- Birds

<u>Consumer Level III</u>

Feeders on larger invertebrates and small fish,

- Large carnivorous fish such as trout, pickerel, bass, sunfish
- Large frogs
- Turtles, snakes
- Birds
- Otter

Bog Community

Under certain conditions, lakes may not develop into ordinary ponds, but into bogs. One notable physical condition of bogs is the scarcity of calcium, and the abundance of organic matter in the water. This latter material may cause the water to appear brown.

The productivity of bog ponds is relatively very low. Aquatic plants, invertebrates and vertebrates may be present, but are limited in species and abundance. The same is usually true of the plankton as well. Around the pond, a mat of vegetation will be projecting over the water. Various shrubs, such as alder and red osier help hold the mat together. Sphagnum is a common moss found in the mat, and may help the bog to develop into a more acid situation.

As succession proceeds in a bog, the mat will eventually creep over the water, until it covers it. Water may still be present under the mat, however. Characteristic bog plants including cranberries, pitcher plants, sundews, bog bean, and several species of sedges, ferns, and orchids will usually be found in this stage. They are able to withstand the physiological drought imposed by the low pH (frequently 4.0 or less) of the bog.

As in the ordinary pond community, the evidence and direction of succession may be observed by noting the zones of vegetation surrounding the bog. The climax community surrounding the bog will eventually encroach upon and occupy the bog site.

FIELD WORK IN THE BOG COMMUNITY

If equipment is available, test the depth of the bog. How long ago did the bog originate? In what way? Is there open water at the center of the bog? What is the bottom of the bog like? If any water exists, obtain a sample and run oxygen and carbon dioxide concentration and pH determinations. Remember that organic materials may seriously affect the pH value obtained.

What is the dominant type of vegetation? What factor is most important in determining the plants which grow here? Make a list of plants found in the bog which are not found elsewhere in your locality. Why are these plants in danger of extinction?

What animals or animal signs can be seen? Do any birds nest in the bog? Are insects abundant? What orders of insects are most common? Look for runways of small mammals in the moss. List the organisms in each trophic level, and their abundance. (See p. 121.)

Is the vegetation of the bog rooted in the bottom? Look for evidence of succession around the edges of the bog. What will eventually be the fate of the bog? How long do you think it will take?

REFERENCES

Aldrich, J. W. 1943. Biological survey of the bogs and swamps in northeastern Ohio. Am. Midland Naturalist 30:346-402.

Dansereau, Pierre, and Fernando Segadavianna. 1952. Ecological study of the peat bogs of eastern North America. I. Structural evolution of the vegetation. Canad. J. Botany 30:490-520.

Gates, F. 1942. The bogs of northern lower Michigan. Ecol. Monogr. 12:213-254.

Grant, M. L. and R. F. Thorne. 1955. Discovery and description of a sphagnum bog in Iowa, with notes on the distribution of bog plants in the state. Proc. Iowa Acad. Sci. 62:197-210.

Heatwole, Harold and Lowell L. Getz. 1960. Studies on the amphibians and reptiles of Mud Lake Bog in southern Michigan. Jack-Pine Warbler 38:107-112.

Kurtz, Herman. 1928. Influence of sphagnum and other mosses on bog reactions. Ecology 9:56-69.

Lindeman, R. L. 1941. The developmental history of Cedar Creek Bog, Minnesota. Am. Midland Naturalist 25:101-112.

Marshall, W. H. and M. F. Buell. 1955. A study of the occurrence of amphibians in relation to bog succession. Itasca State Park, Minnesota. Ecology 36:381-387.

Rigg, G. B. and C. T. Richardson. 1938. Profiles of some sphagnum bogs of the Pacific coast of North America. Ecology 19:408-434.

See ALSO references on pp. 115-116.
For identification manuals, see pp. 218-226.

Swift Stream Community

Where the terrain undergoes a rather rapid change in elevation, streams draining the land are usually rapid-flowing. Such streams gradually erode their streambeds, and, eventually, the entire watershed. In this manner mountains wear down to hills, and hills to plains. As this occurs, the stream flow becomes progressively slower, until when the land is leveled, it is very slow. Small streams may become intermittent, or even dry up in droughts. As this physical change takes place, a concurrent change in the biota of the streams occurs. The changes represent a type of succession in which physical forces are dominant. The swift stream community is the first of the two major types that may be discerned in this succession.

Since small particles are carried downstream more easily than larger ones, the bottom of a swift stream typically consists of rocks or pebbles. The swifter the stream, the larger will be the rocks found. The current allows excellent aeration so that oxygen concentrations are high. Carbon dioxide produced by animals will also be able to escape easily. Temperature of the water will follow closely that of the atmosphere, especially where the water is shallow. However, shading of small, swift streams may prevent heating by the sun.

Organisms are affected by the current and the other physical conditions it imposes. Rooted aquatic plants are unable to survive, only microscopic algae and a few larger non-vascular types being present. Animals must be able to hide under or behind stones, and stay on the bottom. Their bodies may be adapted to this requirement in a variety of ways. Some, like blackfly larvae, have suckers or holdfasts, while others, such as the water penny, the larva of a small aquatic beetle, are much flattened. Many species secrete silklike lines by which they are able to maintain their position. The fishes of swift streams are usually streamlined, and dart rapidly from cover to cover. Even with these protective measures, these organisms are often swept away during high water, and many reach unsuitable places and perish. However, the swift environment offers, for those organisms which have become adapted to it, abundant oxygen, abundant space, and few enemies, so that large numbers may occur in a small area.

FIELD WORK IN THE SWIFT STREAM COMMUNITY

Start the study of this community by determining the temperature, pH and oxygen and CO_2 concentrations of the water. How do they compare with a nearby slow stream? Determine width, depth, rate of flow, and nature of the bottom. Is the water silty and turbid, or clear and clean? Is there evidence of flooding along the stream? Can you determine maximum depth from such evidence? From topographic maps, determine the elevation change in one mile of stream. How does this gradient vary in different sections of the stream?

Examine the study area carefully for vascular plants in the water. Are there microscopic algae or mosses? Collect plankton samples for later analysis in the laboratory. Determine the type of terrestrial community along the shore. What are the dominant species?

Collect a sample of the fishes with a seine of suitable size.

Note adaptations for swift stream life. If the bottom is stony, turn over stones just upstream from the net to collect salamanders and certain large invertebrates such as hellgramites. Take a large stone to shore and examine it for invertebrates. Mark off a quadrat a foot square on such a stone and count the numbers of invertebrates per square foot. Note any adaptations to swift stream life to be observed in the flatworms, hydras, caddisworms, blackfly larvae and other organisms.

If there are deeper areas, pools, backwaters or slower moving sections of the stream, compare the life here with that which you observed in the swifter portion of the stream. In which section is animal life most abundant? Where are the largest animals found? What species is most numerous in each habitat? Is this species herbivorous, carnivorous, or omnivorous? List the numbers and kinds of organisms of each trophic level. (See p. 121.)

From what is known of the geologic history of this region, can you tell if this is an old or young stream? How have its characteristics changed since its origin? In what ways may it change in the future? What will be the effect of such changes upon the organisms of this community? Can you estimate how long these changes may take?

REFERENCES

Dodds, G. S. and F. L. Hisaw. 1924. Ecological studies of aquatic insects. I. Adaptations of mayfly nymphs to swift streams. Ecology 5:137-148.

________________. 1924. Ecological studies of aquatic insects. II. Size of respiratory organs in relation to environmental conditions. Ecology 5:262-271.

________________. 1925. Ecological studies of aquatic insects. III. Adaptation of caddisfly larvae to swift streams. Ecology 6:123-137.

Gumtow, R. B. 1955. An investigation of the periphyton in a riffle of the West Gallatin River, Montana. Trans. Am. Microscop. Soc. 74:278-292.

John, Kenneth R. 1964. Survival of fish in intermittent streams of the Chiricahua Mountains, Arizona. Ecology 45:112-119.

Kuehne, Robert A. 1962. A classification of streams, illustrated by fish distribution in an eastern Kentucky creek. Ecology 43:608-614.

Shelford, V. E. 1911. Ecological succession: stream fishes and the method of physiographic analysis. Biol. Bull. 21:9-34.

Waters, Thomas F. 1961. Standing crop and drift of stream bottom organisms. Ecology 42:532-537.

See ALSO references on pp. 115-116.
For identification manuals, see pp. 218-226.

Slow Stream Community

As elevation gradients of streams decrease, the rate of flow slows. With a slower current, only fine particles are moved downstream, and so the bottom consists of small particles of soil. The slower the current, the more minute will be these particles. Hence, slow streams have bottoms of sand, mud or silt. In general, oxygen is not as abundant as it was in the swift stream, but the carbon dioxide concentration is greater.

Since the current is so reduced, rooted aquatic plants are able to grow along the banks. They will frequently be found entirely across a shallow stream. In such cases they may act as a strainer to stop the passage of plankton farther down the stream. Plankton found here will have a different composition from that of swift streams. Species of aquatic plants in this community are similar to those found along the shores of eutrophic lakes and ponds.

Animals are more varied in this community than they were in the swift stream community. Fish are more numerous, but a different group of organisms is found on the bottom. Instead of streamlined forms or those adapted for clinging, animals found here are adapted for burrowing in the mud. Usually, animals of this habitat are more tolerant of reduced oxygen concentration and warmer temperatures.

Large slow streams may become relatively shallow, and may periodically flood their valleys. Smaller slow streams may dry up in periods of drought, leaving only puddles that represent the deeper parts of the stream. Organisms that are to be successful in this circumstance must be able to survive these periods when oxygen is very low and temperatures high.

FIELD WORK IN THE SLOW STREAM COMMUNITY

Before the water is disturbed, collect water samples for oxygen, carbon dioxide and pH determination. Is the water clear or muddy? Note the width and maximum depth of the stream. What kind of bottom materials are present? Is there any indication of maximum height of the stream during floods? From topographic maps, determine the elevation gradient (feet of elevation change per mile). How does it compare with that of the swift stream? What kinds of vascular plants are present? Under what circumstances might a slow stream be devoid of rooted aquatic plants? Is there any indication of zonation of plant communities? Collect plankton samples to be examined in the laboratory.

Seine the stream along the shore, including deeper holes where larger fish may hide. Search the aquatic vegetation for animals, and strain the mud for invertebrates. Mollusks may be found in relatively deep water off shore in sandy areas. How many of the invertebrate species are adapted for life in mud? How many utilize vegetation for maintaining their position? Are there more species of invertebrates here than in the swift stream? Vertebrates? How does the total number of animals compare in the two communities? List the number and kinds of organisms of each trophic level. (See p. 121.)

What is the past geologic history of this region of the stream? What may its future development be? How will this affect the kind of organisms present?

REFERENCES

Abel, D. L. 1959. Observations on mosquito populations of an intermittent stream in California. Ecology 40:186-193.

Berner, L. M. 1951. Limnology of the lower Missouri River. Ecology 32:1-12.

Blum, J. L. 1956. The ecology of river algae. Botan. Rev. 22:291-341.

Gersbacher, Willard M. 1937. Development of stream communities in Illinois. Ecology 18:359-390.

Greenberg, Arnold E. 1964. Plankton of the Sacramento River. Ecology 45:40-49.

Larimore, R. Weldon, William F. Childers and Carlton Heckrotte. 1959. Destruction and re-establishment of stream fish and invertebrates affected by drought. Trans. Am. Fisheries Soc. 88:261-285.

Shelford, V. E. and S. Eddy. 1921. Methods for the study of stream communities. Ecology 10:382-391.

Stehr, William C. and J. W. Branson. 1938. An ecological study of an intermittent stream. Ecology 19:294-310.

See ALSO references on pp. 115-116.

For identification manuals see pp. 218-226.

Marine Shore Communities

The edge of the sea is an area of special fascination for the field biologist. It represents the point at which organisms first left the relatively stable conditions of life in the ocean, to enter the more varied but more rigorous conditions of terrestrial life. In the region between low tide and the splash zone (see Fig. 11), plants and animals can exist only if they have evolved the means to circumvent or overcome the special difficulties of the habitat. These difficulties include increased light intensity, respiration out of water, variability in salinity and temperature, alternate flooding and dessication, wave shock, abrasion, and problems associated with feeding and being fed upon.

The single physical factor which is probably most important in determining the types of organisms inhabiting an area of shoreline is exposure: whether the shore is totally exposed to wave action, partially exposed, or completely protected. Second only to exposure as a physical factor is type of substrate: whether the beach is rocky, sandy, or muddy. These two factors are not totally unrelated. On exposed shores, rock cliffs and rocky beaches are to be expected, but muddy beaches are non-existent. On fully protected shores, on the other hand, muddy beaches are common and rocky beaches rare.

The firmness of the substrate is important because of its function as a place of attachment for organisms. Individuals which are firmly attached to rock may hold their place under the impact of waves, or of the shifting and moving tides. Mud gives the least support to an organism, and the combination of exposure and substrate thus affects the species composition of seashore communities.

On rocky beaches there is a zonation of algae from low tide through the strand line (see Fig. 11),which results primarily from variations in light intensity. The splash zone is characterized by blue-green algae (Cyanophyta), lichens and other dark-colored plants which have resulted in the name, black zone. Below this, near high tide line, is the tan or yellow zone, so called because of the color of the predominant plants; then the green zone, dominated by green algae (Chlorophyta), and the brown zone, where brown algae (Phaeophyta) give the typical appearance to the lower intertidal zone. Red algae (Rhodophyta) are found in the constantly submerged (subtidal) areas. On sandy or muddy beaches, where the substrate is not stable enough to permit the growth of algae, this color banding is absent.

The diagrams and descriptions which follow are generalized, although the rocky beach forms a basis from which to begin. In every case, variations of physical factors will provide variations in the communities, so that a comparison of different communities, and of the factors responsible for their formation, may be made readily.

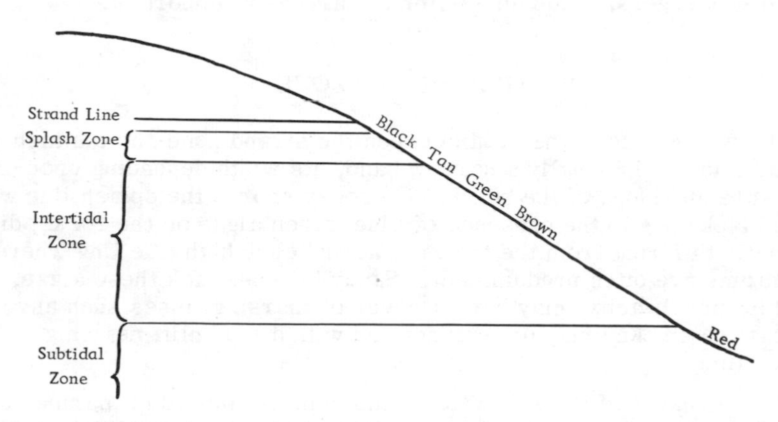

FIGURE 11. ZONATION OF ALGAE ON ROCKY SEA SHORES

THE STRAND LINE

The strand line is the beach zone where debris and flotsam are left as the tide advances and recedes. It is usually narrow, and contains such dead organisms as fishes, invertebrates and algae, as well as light bulbs, cans, bottles and driftwood. Sandy beaches have especially rich strand lines and are particularly attractive to beachcombers, shell collectors and the like. Mud beaches have such a wide area of strand that any line is scarcely distinguishable as such. This is especially true where mud beaches are covered with close-growing vegetation such as salt meadow cordgrass (Spartina sp.) or an ecologically similar species.

Both terrestrial and marine animals inhabit the strand line. Some are scavengers on the dead organic matter, some are predators on the scavengers, and some are both. Marine fauna utilizing this zone must be able to obtain oxygen from the air instead of water (as in snails which have a modified mantle) or to hold water in special branchial chambers (as in certain crabs). These marine forms are also usually secretive, burrowing in the soil or hiding in the debris, thus achieving some degree of protection from predators such as beetles, flies, seagulls, rats, skunks and sandpipers. Some are nocturnal, while others avoid enemies through swiftness of foot. Another severe difficulty for inhabitants of the strand line is exposure to conditions of weather and climate - heat and dessication in the summer, freezing in winter.

Arthropods make up the greatest number of strand line organisms, and include crabs, beach fleas, sow bugs, insects and arachnids. Among the terrestrial visitors which act as both predators and scavengers, birds and mammals are most important.

THE SPLASH ZONE

The splash zone is the area between the strand zone and the high tide line. It is usually a narrow band, its width depending upon exposure and slope of the beach. On rocky shores the splash line will be black, due to the presence of blue-green algae on the rocks, distinctly different from the tan area at or below high tide line where diatoms are often predominant. Sandy beaches lack these algae, while mud beaches may have a cover of marsh grasses such as *Spartina* sp. and may or may not end with a mud cliff near high tide line.

Because of the salt water, this zone is limited to marine animals except for occasional visitors from terrestrial habitats during low tides. On rocky beaches, the lack of protective cover in addition to the fluctuations in the presence of salt water requires inhabitants of the area to feed rapidly. These animals are exposed to rain and dessication during low tide, thus presenting a salinity problem as well as a water maintenance problem and a respiratory problem. Temperatures in this zone vary greatly during the day. Some species close up during the day and cool themselves by respiration, becoming active only at night. Seasonal composition of the fauna of this zone may vary also, some species leaving it in winter for the intertidal or subtidal zones. Predation by terrestrial forms is a constant danger between tides, and the marine forms are usually strongly armored and well attached to the substrate in rocky areas. On sandy or mud beaches most animals burrow into the substrate.

Food, in the form of attached algae, plankton (during blooms) and organic detritus, is abundant in the splash zone of rocky beaches. Sandy beaches have few inhabitants because of the unstable substrate, while abrasion on gravel beaches limits certain organisms. Animals of the splash zone may include snails (*Littorina* sp.) which are confined to this zone, especially on rocky beaches. Limpets, beach fleas, barnacles, crabs, and isopods are common. In addition to the blue green algae, diatoms intermix in the plant species, especially near high tide line. Sandy and muddy beach splash zone inhabitants may include tiger beetles and other beetles, spiders, flies, earwigs, ghost crabs, and beach amphipods.

INTERTIDAL ZONE

Within the intertidal zone there is a region that receives water nearly all the time (except at low tide), gradually changing to a region that is submerged only for minutes at a time (at high tide). This condition, in addition to variations in exposure, substrate type, and other conditions, allows a wide diversity of habitats and causes environmental instability. As a result, there is an amazing variety of inhabitants in the intertidal zone. These organisms are exposed to a great deal of light and so are often light-colored. Shade is afforded to some sessile animals by other small animals living on them. On mud and sand beaches, burrowing may give relief from light and heat, while on rock beaches, crevices, stones, and tide pools may provide shelter. There seems to be a tendency for animals to seek shelter to avoid temperature extremes as well as unfavorable light.

Oxygen is not a limiting factor for organisms on rocky beaches. Some animals have adapted to this area by means of unusually large gill filaments. On sandy beaches, spaces between sand grains hold water and dissolved oxygen, but on mud beaches breathing tubes must be manufactured, since the mud itself is lacking in oxygen and is high in hydrogen disulfide from bacterial decay of organic matter.

Salinity is variable with location, season, and climate (because of rain and shore seepage), and inhabitants must be widely tolerant of this factor.

Food is more abundant in this zone than in any other area of the sea, per unit volume of water. Not only are there more plankton algae of various kinds, but attached algae and bacteria are also numerous. Large herbivores, feeding on algae, include periwinkles, limpets, isopods, crabs and fishes. Plankton provides food for barnacles and bivalve molluscs, the most characteristic animals of an exposed rocky beach. Sea urchins and amphipods act as scavengers, feeding on dead algae. Smaller particulate organic matter is the mainstay of many gastropods, echiuroids, brachiopods, bryozoans, tunicates, worms and sand dollars. Bacteria may be the main source of food for some of the worms which live in mudbeaches.

With this variety of food, and animals feeding on it, it is not surprising that predators are more abundant here than in any other zone. Common predators of the intertidal zone include starfishes, gastropods, nemertines, sea anemones, fishes and birds. Various means of protection against predation and the rigors of the surf are found in these intertidal organisms. Many have strong skeletons, others are firmly attached by unique devices, while still others burrow in sand or mud or even in rock. Some gain protection through camouflage, while a few specialize in a speedy getaway. Predators, however, have equally good adaptations for their way of life. Oyster drills can drill their way into an oyster, and moon shells burrow as readily as the clams which are their prey. Even the sturdiest bivalve is not proof against the gulls, which drop these molluscs from a considerable height onto rocks. Hard-surfaced roads along beaches are good places to observe this phenomenon.

Wave action is a dominating physical factor in most intertidal zones. Since water moves up and down the beach once or twice a day, any one area is exposed to waves or breakers two or four times in twenty-four hours. Tidal currents render attachment mechanisms necessary, such as those found among the molluscs. The boring species, such as angel's wings, may use this means to escape from currents as well as from predators. Among the species which bore, a few even imprison themselves in rock.

Sandy beaches are least productive of intertidal habitats because of the unstable substrate, wave shock, and abrasion from moving sand. Dessication is not severe because the size of the sand particles allows capillarity to bring water above sea level, and burrowers can reach damp sand or the water level. On such beaches burrowing is the primary means of avoiding wave action, although some species, such as the mole crab, escape heavy waves by moving up and down the beach ahead of the breakers. Burrowing bivalves and arthropods are predominant on sandy beaches, while annelid worms, shrimps, burrowing sea urchins and sea anemones may be found here.

Mud beaches form only in protected places, so that wave action and abrasion are not severe problems on such beaches. Although burrows in mud must usually be supported by tubes, some forms, such as the predaceous nemertines, form no permanent burrows as they wander through the mud. Many of the herbivores here feed almost entirely on bacteria. Clams, feeding on dead organic matter, are frequently abundant. Tube-forming worms are the other dominant organisms of this community.

Tide pools often occur in the intertidal zone, and within them are to be found most of the animals that inhabit the zone. Because the upper and lower ends of the intertidal zone represent extremes of environmental conditions, the intertidal pool organisms will differ from the upper to the lower beaches. Near the splash zone green algae, mussels, barnacles, and starfish are dominant, while snails, crabs, limpets and brittle stars also share this area. In contrast, the region next to low tide mark is dominated by sponges, tunicates, nudibranchs, chitons, hydroids and sea urchins. Shrimps, scallops, burrowing clams and nemertines also occur here. Various fishes, such as the blenny, occupy the intertidal zone. Plants of the zone include brown algae, especially _Fucus_ spp.

Among the problems of survival for tidepool inhabitants are variation in salinity and temperature, especially in shallow pools. Rain, and evaporation from exposure to the sun may decrease or increase salinity, respectively. If pools are subjected to a change in one of these factors over a long period of time (as in a pool near the splash zone) it can cause the death of many individuals of various species. Time seems to be even more important than severity of the variation. As the temperature of the water increases, it is unable to maintain dissolved oxygen, and this too can result in a lethal situation.

FIELD WORK IN MARINE SHORE COMMUNITIES

If possible, visit the seashore during periods of low tide. Tide tables are available from the Superintendent of Documents, Government Printing Office, Washington, D. C. Exposed beaches especially can be studied for any great length of time only during low tide. Study the part of the beach which will be exposed the shortest period first. If you arrive at low tide, study the region next to the water first, working up toward shore.

Each community should be examined thoroughly. Dip nets, shovels and wire screens will help you gather animals in tide pools or in sand or mud. Geologist's hammers are useful to extricate animals from rocky crevices. Bait such as the flesh of clams may be the means of luring animals from hiding places in tide pools. Knives are necessary to cut into the holdfasts of large algae where many animals may hide. In rocky areas stones should be carefully turned (then replaced) to detect animals. The entire surface of algae, either that which is living and attached, or that which is stranded, should be examined. The strand line may be very rich in life, especially in the soil under it. In sandy areas of all zones, dig especially where there is any sign of burrows or tunnels.

Measurements of the environment may be made. Temperatures of the substrate, water in tidepools at various levels, and in the sea may be of significance. The oxygen tension and pH of the water of the tide pools might be obtained. Observe the amount of dessication of the substrate at different levels of the intertidal zone, and in the splash zone, and strand line. Note whether there is moisture under rocks or seaweed.

An estimate of numbers of organisms present (at least the larger ones) may be made by means of quadrats. In such analyses, the populations on the surface might be compared to those under rocks and seaweed, or in the sand or mud to a certain depth. Adjust the size of the quadrat to suit the situation.

While collecting animals and data, be certain to be precise as to the exact location of the collection on the beach. Any animals that cannot be identified on the spot should be preserved and complete collection data should be included with each specimen.

QUESTIONS

1. In each zone, which plants and animals are characteristic?
2. Which of these plants or animals might be considered to be terrestrial in origin? Which are of marine origin and adapted to terrestrial existence?
3. Which species are plant scavengers, animal scavengers, producers, predators, grazers? Which were the dominant ones in each zone?
4. What special adaptations do you find in these animals with regard to:

   a. color
   b. movement
   c. period of activity
   d. respiration
   e. food getting
   f. reproduction
   g. protection from enemies

5. Which of the species reproduce in their zone, which migrate for the purpose?
6. If beaches exhibiting more than one type of exposure to waves or substrate type were visited, compare the physical conditions and organisms found in each zone.
7. Was any vertical distribution of any gẽnera obvious? (Especially seen in periwinkles and limpets.) Which of the species you encountered had the widest intertidal range?

REFERENCES

Benton, Allen H. and William E. Werner, Jr. 1965. Field Biology and Ecology. 2nd Ed. McGraw-Hill Book Co., N. Y.

Buzzati-Traverso, A. A. 1958. Perspectives in marine biology. Univ. of Calif. Press, Berkeley, Calif.

Carson, Rachel. 1959. The edge of the sea. New Amer. Lib., New York.

Chapman, V. J. 1957. Marine algal ecology. Botan. Rev. 23:320-350.

Coker, R. E. 1947. This great and wide sea. Univ. of North Carolina Press, Chapel Hill, N. C.

Crowder, William. 1931. Between the tides. Dodd, Mead & Co., N. Y.

Dahl, Erik. 1953. Some aspects of the ecology and zonation of the fauna of sandy beaches. Oikos 4:1-27.

Doty, M. S. 1946. Critical tide factors that are correlated with vertical distribution of marine algae and other organisms along the Pacific coast. Ecology 27:315-328.

Hedgpeth, Joel W. (ed.). 1957. Treatise on marine ecology and paleoecology. Vol. I. Ecology. Geol. Soc. Mem. 67. Geol. Soc. Amer., N. Y.

Hewatt, Willis G. 1937. Ecological studies on selected marine intertidal communities of Monterey Bay, California. Am. Midland Naturalist 18:161-206.
Klugh, A. Brooker. 1924. Factors controlling the biota of tide pools. Ecology 5:192-196.
Lewis, J. R. 1964. The ecology of rocky shores. English Univ. Press, London.
Light, S. F., R. I. Smith, F. A. Pitelka, P. Abbott, F. M. Weesner. 1954. Intertidal invertebrates of the central California coast. Univ. of Calif. Press, Berkeley, Calif.
MacGinitie, G. E. 1939. Littoral marine communities. Am. Midland Naturalist 21:28-55.
__________ and N. MacGinitie. 1949. The natural history of marine animals. McGraw-Hill Book Co., N. Y.
Moore, Hilary B. 1958. Marine ecology. John Wiley, N. Y.
Pearse, A. S., H. J. Humm, and G. W. Wharton. 1942. Ecology of sand beaches at Beaufort, N. C. Ecol. Monogr. 12:136-190.
Ricketts, E. F. and J. Calvin. 1952. Between Pacific tides. Third Ed. Stanford Univ. Press, Stanford.
Riley, Gordon A. 1963. Marine biology. I. Amer. Inst. Biol. Sci., Washington.
Russel, F. S. (ed.) 1963. Advances in marine biology. Academic Press.
Sears, Mary (ed.) 1961. Oceanography. Am. Assoc. Adv. Sci. Washington.
Stephenson, T. A. and A. Stephenson. 1949. The universal features of zonation between tide-marks on rocky coasts. J. Ecology 37:289-305.
Yonge, C. M. 1949. The seashore. Collins, London.
Zobell, C. E. and C. B. Feltham. 1942. The bacterial flora of a marine mud flat as an ecological factor. Ecology 23:69-78.
For identification manuals, see pp. 218-226.

Primary Productivity

The primary productivity of an aquatic community is the rate at which energy from the environment is converted into organic compounds (foods) by photosynthetic and chemosynthetic organisms. This productivity may be measured by several techniques. One method frequently used in aquatic communities is measurement of oxygen output of the producers, primarily algae and other aquatic plants. The oxygen produced over a given time period is used as an index of productivity. This is possible because of the relationship between oxygen production and carbohydrate production. Thus, in the overall photosynthetic process,

$$6\ CO_2 + 6\ H_2O \xrightarrow[\text{chlorophyll}]{\text{light}} C_6H_{12}O_6 + 6\ O_2$$

In other words, for each molecule of a six-carbon carbohydrate produced, 6 molecules of oxygen are released. Of course, oxygen is simultaneously being consumed by respiration. Also, it must be remembered that only in the presence of light can photosynthesis occur. Thus there will be a diurnal cycle of oxygen production. Furthermore, light intensity will vary with time of day, extent of cloud cover, condition of the water surface (choppy water interferes with light transmission), and season of the year. In addition, depth of penetration of light will be affected by turbidity caused by sediment or plankton blooms, and these conditions may vary from time to time. Temperature of the water will affect all metabolic processes, and the temperature will of course vary with depth and with the seasons.

To measure total oxygen productivity, the oxygen content of water from desired locations and depths is determined at the beginning of the experiment. Then two bottles of water are taken from each of these locations and depths, the bottles sealed and one of each pair is covered with aluminum foil to exclude all light. In the uncovered bottle, photosynthesis as well as respiration of plants, animals and bacteria present will occur. In the covered bottle, only respiration of the organisms will proceed. All three samples of water should be taken from the same spot (using sampling precautions necessary for making a Winkler oxygen determination, see p. 67) and two of the three samples, the covered and uncovered sealed bottles, should be returned to the spot from which the samples were taken. The water in these bottles should include photosynthetic organisms, such as phytoplankton or larger plants.

At the end of a given time period, each of the bottles is analyzed for oxygen content. (The time period may be one to several hours, up to 24.) Total oxygen production for the time interval will be sum of the differences between the oxygen content at the start of the experiment and the oxygen content in each bottle at the end. For example:

Oxygen at end, bottle not covered		11.1 ppm O_2
Oxygen at start	-	7.2
Net oxygen produced		3.2 ppm O_2
Oxygen at start		7.2 ppm O_2
Oxygen at end, bottle covered	-	2.6
Oxygen consumed		4.6 ppm O_2
Net oxygen produced		3.2 ppm O_2
Oxygen consumed	+	4.6
Total oxygen produced		7.8 ppm O_2

PROCEDURE

If unicellular algae are common, samples of water will include sufficient specimens to produce satisfactory results. In winter time, such plants may be at a minimum, but the technique may be demonstrated by using sprigs of Elodea or other aquarium plants.

1. Take a sample of water from each of the assigned locations and depths.
2. Run a Winkler test on each of these samples (see p. 67).
3. From each of the same locations and depths indicated above, obtain 2 more samples. If using Elodea or similar plants, place sprigs of equal size in each bottle.
4. Seal the bottles with paraffin.
5. Cover one bottle of each pair with aluminum foil so as to exclude all light.
6. Suspend each pair of bottles (one covered, one uncovered) in the location from which the water sample within was taken.
7. At the end of the testing period, remove both bottles from the water. Invert the bottles to mix any bubbles in the water.
8. If Elodea or large plants were used, remove the sprigs and replace the lost volume with water from the sample location. (This will be a source of error, but should not be significant.)
9. Quickly run the Winkler test on all samples.
10. Calculate the total oxygen produced.

If a pair of light and dark bottles were placed 1/2 meter below the surface and at intervals of one meter to the bottom, the productivity of a column of water at that particular time and location would be sampled. If the sodium thiosulfate solution is standardized so that 10 cc. of it will neutralize 10 cc. of potassium dichromate, then the oxygen values in ppm. multiplied by 0.698 will give the oxygen production in cubic centimeters per liter. (See Welch pp. 208-209 for standardization procedure.) Since there are 1000 liters per cubic meter, multiplying the value in cc./L by 1000 will give the cc. of oxygen dissolved in a cubic meter. By adding the oxygen produced at each depth of one location, we may find the total oxygen productivity for a "core" of water one meter square at the surface. In one experiment in the Gulf of Maine (Clark and Oster, 1939) 29.1 liters (29,100 cc.) of oxygen were produced in 9 hours 10 minutes (depth sampled was 50 meters, the compensation depth was 27 M).

REFERENCES

Clark, G. L. and R. H. Ostler. 1934. The penetration of the blue and red components of daylight into Atlantic coastal waters and its relation to phytoplankton metabolism. Biol. Bull. 67:59-75.

Dugdale, R. C., and J. T. Wallace. 1960. Light and dark bottle experiments in Alaska. Limnol. Oceanogr. 5:230-231.

Edmondson, W. T. 1956. The relation of photosynthesis by phytoplankton to light in lakes. Ecology 37:161-174.

Goldman, Charles R. and Robert G. Wetzel. 1963. A study of the primary productivity of Clear Lake, Lake County, Calif. Ecology 44:283-294.

Odum, Eugene. 1959. Fundamentals of ecology. W. B. Saunders Co., Phil., Pa.

Verduin, Jacob. 1956. Primary production in lakes. Limnol. Oceanogr. 1:85-91.

______________. 1959. Photosynthesis by aquatic communities in northwestern Ohio. Ecology 40:377-383.

Welch, Paul S. 1948. Limnological methods. The Blakiston Co., Phil., Pa.

AQUATIC DATA SHEET

STUDENT:____________________ DATE:_______________

State:____________________ County:____________________

Locality:____

Drainage:____

Weather:_______________ Air temp.:________ Water pH:________

SITE A	SITE B	SITE C
Locality: ______ Water depth: ______ Water temp.: ______ O_2 conc.: ______ CO_2 conc.: ______	______ ______ ______ ______ ______	______ ______ ______ ______ ______
Species	Species	Species

AQUATIC DATA SHEET

STUDENT: ______________________ DATE: ______________

State: ____________________ County: ________________

Locality: ____

Drainage: ____

Weather: ______________ Air temp.: _______ Water pH: ______

SITE A	SITE B	SITE C
Locality: ________	________	________
Water depth: ________	________	________
Water temp.: ________	________	________
O_2 conc.: ________	________	________
CO_2 conc.: ________	________	________
Species	Species	Species

AQUATIC DATA SHEET

STUDENT:______________________ DATE:______________

State:____________________ County:__________________

Locality:____

Drainage:____

Weather:______________ Air temp.:______ Water pH:______

SITE A	SITE B	SITE C
Locality: ________	________	________
Water depth: ________	________	________
Water temp.: ________	________	________
O_2 conc.: ________	________	________
CO_2 conc.: ________	________	________
Species	Species	Species

AQUATIC DATA SHEET

STUDENT: ______________________ DATE: ______________

State: ____________________ County: ____________________

Locality: ____

Drainage: ____

Weather: ______________ Air temp.: ________ Water pH: ________

SITE A	SITE B	SITE C
Locality: ________	________	________
Water depth: ________	________	________
Water temp.: ________	________	________
O_2 conc.: ________	________	________
CO_2 conc.: ________	________	________
Species	Species	Species

AQUATIC DATA SHEET

STUDENT:______________________________ DATE:____________________

State:____________________________ County:__________________________

Locality:____

Drainage:____

Weather:____________________ Air temp.:________ Water pH:________

SITE A		SITE B	SITE C
Locality:	________	________	________
Water depth:	________	________	________
Water temp.:	________	________	________
O_2 conc.:	________	________	________
CO_2 conc.:	________	________	________
Species		Species	Species

Section V
Population Studies

Introduction

In modern biology, the study of populations as well as of individuals has assumed great importance. The study of populations can give us much information about the fundamental biology of the species which cannot be learned by study of the individual. We may learn the actual population at any given time; the biotic potential of the species; the age and sex composition of animal populations; the growth rate of trees; the fluctuations which occur in populations at different times. We may learn the weak points in the life history of pest species which we wish to reduce in numbers, or how to maintain adequate numbers of desirable species.

Population studies of various kinds of animals require the use of different techniques. Mammals are most frequently studied by trapping, and live traps are preferred because they do not alter the population while the study is in progress. Birds are sometimes trapped, but more frequently bird populations are studied by direct observation. Reptiles and amphibians which are of suitable size and habits may also be observed directly, while fishes and other less readily observable forms are studied by one of several indirect methods. Invertebrates may also be studied using techniques similar to those used for vertebrates or with specialized methods. Several of the techniques previously described are essential for a population study. See especially references on marking techniques, recognition of animal signs, trapping techniques, and collection methods in Section II.

The following exercises are intended to introduce you to the techniques of population studies in a relatively simplified manner suitable for use when limited time is available. They will give you practice in using different methods for arriving at similar data in different animal groups.

REFERENCES

Andrewartha, H. G. 1961. Introduction to the study of animal populations. Univ. of Chicago Press, Chicago, Ill.
__________, and L. C. Birch. 1954. The distribution and abundance of animals. Univ. of Chicago Press, Chicago, Ill.
Elton, Charles. 1942. Voles, mice and lemmings: problems in population dynamics. Oxford Univ. Press, London.
__________. 1958. The ecology of invasions by animals and plants. Methuen & Co., London.
Fisher, R. A., A. Steven Corbet, and C. B. Williams. 1943. The relation between the number of species and the number of individuals in a random sample of an animal population. J. Animal Ecology 12:42-58.
Hazen, William E. 1964. Readings in population and community ecology. W. B. Saunders Co., Phila., Pa.
Keith, Lloyd B. 1962. Wildlife's ten-year cycle. Univ. of Wisc. Press, Madison, Wisc.
Lack, David. 1954. The natural regulation of animal numbers. Oxford Univ. Press, N. Y.
Slobodkin, Lawrence B. 1961. Growth and regulation of animal populations. Holt, Rinehart and Winston, N. Y.
For films on population studies, see p. 230.

Field Work on Mammal Populations

METHOD A: USE OF TRAPS

An open field with a relatively large population of small mammals will be chosen by your instructor. Twenty-five trapping stations, of three traps each, will be set at 42 foot intervals in a line across the field. If traps and time are available, several such traplines should be set, with 200 foot intervals between the traplines. Traps may also be set in a grid (Fig. 13, p. 167) with the interval between traps 42 feet. If live traps are used, they should be checked at 12-hour intervals, and each mammal removed, marked, its data recorded, and released at or near the point of capture. If snap traps are used, they may be checked at 24-hour intervals. Sprung traps should be reset and all mammals taken to the laboratory for study. After three 24 hour periods of trapping, the traps will be taken up and the total catch then tabulated. Where live traps are used, record the retrapped individuals as well as the total new individuals taken. See p. 36 for directions for baiting and setting traps. Unless traps are set in runways or other favored locations, few if any mammals may be caught.

MAMMAL POPULATION
DATA SHEET
PART A

STUDENT:____

Locality:____________________ Dates of study:____________________

No. of traps used:_______________ Habitat:____________________

Weather:____

Date	Name of mammal	Age and Sex	Breeding condition	Habitat

Remarks: (nests, disease, molting, etc.)

MAMMAL POPULATION DATA SHEET PART B

STUDENT: ____________________

Locality: ____________________ Dates of study: ____________________

No. of boards used: __________ Habitat: ____________________

Weather: ____________________ Board Interval: ____________________

Date	Name of mammal represented by scats	Row No.	Board No.	Habitat

METHOD B: USE OF SCAT BOARDS

If trapping is not feasible, or if supplementary information is desired, scat boards may be used. Small boards, 4" x 4", are cut from exterior grade plywood or masonite. These boards are distributed in a line or uniform grid pattern through the sample area, similar to the manner suggested for method A. Small mammals will use these boards for defecation sites, and thus leave a sign of their presence. The frequency of occurrence of scats will give an indication of population size. Population fluctuations, local or ecological distribution, activity rhythms, effects of weather on activity and effectiveness of baits may be studied with this technique. At least 100 boards should be used, and should be visited each 24-hour period for collection of scats. They should be left in place for at least three consecutive days.

The usefulness of scatboards has been increased by development of a technique using dyes to detect movements within the population. A non-toxic dye, such as Fluorescin, Fast Green FCF, or Orange I, is placed on a bait such as rolled oats. By using different colored baits in different parts of the habitat, the movements of mammals around a point may be ascertained. By feeding dyed bait to a single captured animal and releasing him, then collecting scats, the range of movement of a single individual may be determined. (See New, 1958, Haresign, 1960.)

ANALYSIS OF RESULTS

In the area you studied, what was the estimated population per acre of each species collected? (Each 25 trap stations placed at 42 foot intervals, will sample approximately an acre, whether the stations are in a line or a grid.) What was the total population of all small mammals per acre? Which species are predators, which are herbivores, which are omnivores? (Stomachs may be saved for food content analysis.) Did you find any evidence of disease or parasites? Were any pregnant females taken? If so, how many embryos were found? Were males in breeding condition? Was there evidence of molt? What percentage of the specimens taken were immature? Was there evidence of recent parturition? What was the favored habitat of each species trapped? Would you say that the population is high, medium or low?

If live-trapping was employed, calculate the population by means of the Lincoln-Petersen Index (see p. 74). During which 12-hour period were more mammals taken? Were there differences between different species in this respect? Could you determine anything about reproduction in the mammals by external examination?

Were any diseased or parasitized individuals seen? If so, of what type were the parasites? Did you retrap any individual often enough to be able to calculate its home range? List the species in order of abundance. Keep the data from the whole class for future comparison with results of other classes.

If scat boards were used, what percentage of the boards were used daily? Could you detect scats produced by various species? (Animals may be live trapped in order to obtain sample scats for diagnosis of scats collected in the field.) Could you detect a difference in the size of the populations of the various species present? Which was most abundant? What would be the estimate of the population of all mammals present, assuming each mammal visited no more than one scat board in 24 hours? The estimate of the population of each species? If mammals were live trapped, fed dyed food and released, what was the calculated home range of each species so tested? How could such information be used in estimated populations by the scat board technique?

REFERENCES

Allen, D. L. 1950. The purposes of mammal population studies. J. Mammal. 30:18-21.

Bendell, James F. 1961. Some factors affecting the habitat selection of the white-footed mouse. Canad. Field. Nat. 75:244-255.

Cockrum, E. L. 1962. Laboratory and field manual introduction to mammalogy. Ronald Press, New York.

Davis, David E. 1956. Manual for analysis of rodent populations. Edwards Bros., Ann Arbor, Mich.

Dice, Lee R. 1941. Methods for estimating populations of animals. J. Wildl. Mgmt. 5:398-407.

Emlen, John T., Jr., Ruth L. Hine, William A. Fuller, and Pablo Alfonso. 1957. Dropping boards for population studies of small mammals. J. Wildl. Mgmt. 21:300-314.

Getz, Lowell. 1961. Factors influencing the local distribution of _Microtus_ and _Synaptomys_ in southern Michigan. Ecology 42:110-119.

Goertz, John W. 1964. The influence of habitat quality upon density of cotton rat populations. Ecol. Monogr. 34:359-381.

Haresign, Thomas. 1960. A technique for increasing the time of dye retention in small mammals. J. Mammal. 41:528.

Hayne, Don B. 1949. Two methods for estimating population from trapping records. J. Mammal. 39:399-411.

MacMillen, Richard E. 1964. Population ecology, water relations, and social behavior of a southern California semi-desert rodent fauna. Vol. 71. Univ. of California Pub. in Zoology.

Mayer, William, and Richard van Gelder (eds.) 1963. Physiological mammology. Vol. I. Mammalian populations. Academic Press, N. Y.

New, John. 1958. Dyes for studying the movement of small mammals. J. Mammal. 41:416-429.

Stickel, Lucille F. and Oscar Warbach. 1960. Small mammal populations of a Maryland woodlot 1949-1954. Ecology 41:269-296.

See ALSO references on p. 150.
For identification manuals, see pp. 218-226.

Field Work on Breeding Bird Populations

An area should be selected which contains a fairly large number of breeding birds. A map should be made of the area, preferably using a topographic map as a base. If there are variations in plant community types in the area, these should be noted on the map, including the dominant species and exact location of each community. If time permits, a census grid may be set up (see p. 165) and regular surveys made over a period of time. Since an area of 209 x 209 feet is approximately one acre, by setting up parallel lines 209 feet apart, students walking along such lines will be able to survey one acre for every 209 feet of their line. Each student is given a map, and on it notes all the birds observed, their nesting or other behavior (see below). As large an area as is feasible should be censused, up to about 50 acres. Similar results may be obtained by mapping the area and placing each student at a designated spot with a copy of the map. If each observation spot is approximately 209 feet from every other spot in the form of a grid, each student would thereby census one acre. If students are not well versed in bird identification, a single common species may be studied.

Things to be observed and noted on the map or data sheet include the following:

1. numbers of each species
2. sex of each individual
3. nests
4. numbers of eggs in nest
5. behavior - singing; feeding; courtship activity; nesting; intraspecific strife; interspecific strife; other

After the period of observation is completed, data from all maps is entered on a master map.

What do you estimate to be the total population (in breeding pairs) of birds in this area? How many pairs per acre does this represent? Can you see evidence of habitat preference from your data? Do any pairs seem to have a well-established territory? Was there any evidence of intra-specific strife or territorial defence? Were any observations made on courtship, nesting, feeding, or other aspects of behavior? Were any nests located? Were any species colonial in distribution?

STANDARD SIGNS FOR USE IN BIRD POPULATION STUDIES

x = singing male

• = non-singing bird

(4) = nest with 4 eggs

y = young

⟶ = direction of flight

Abbreviations for each species may be developed to suit the bird life of the locality: e.g.,
chsp = chipping sparrow, chsw = chestnut sided warbler, etc.

REFERENCES

Aldrich, John W. 1947. Counting the birds. Audubon Mag. 49:214-220.

Beak, Edward. 1960. Forest bird communities in the Apostle Islands of Wisconsin. Wilson Bull. 72:156-181.

Beer, J. R., L. D. Frenzel and N. Hansen. 1956. Minimum space requirements of some nesting Passerine birds. Wilson Bull. 68: 200-209.

Brewer, Richard. 1963. Stability in bird populations. Occ. Pap. C. C. Adams Center for Ecol. Stud. No. 7.

Graber, Richard R. and Jean W. Graber. 1963. A comparative study of bird populations in Illinois 1906-1909 and 1956-1958. Illinois Nat. Hist. Survey Bull. 28(3):383-528.

Hagar, Donald C. 1960. The interrelationships of logging, birds, and timber regeneration in the douglas-fir region of northwestern California. Ecology 41:116-125.

Hutson, H. P. 1956. The ornithologists guide. Philosophical Library, N. Y.

Johnston, Richard F. 1961. Population movements of birds. Condor 63:386-389.

Kendeigh, S. C. 1944. Measurement of bird populations. Ecol. Monogr. 14:67-106.

Martin, N. D. 1960. An analysis of bird populations in relation to forest succession in Algonquin Provincial Park, Ontario. Ecology 41:126-140.

Odum, E. P. and E. J. Kuenzler. 1955. Measurement of territory and home range size in birds. Auk 72:128-137.

Pettingill, Olin S. 1956. A laboratory and field manual of ornithology. Burgess Publ. Co., Minneapolis.

Pough, Richard. 1947. How to take a bird census. Audubon Mag. 49:288-297.

For identification manuals, see pp. 224-225.
For identification records, see p. 230.
See ALSO references on p. 150.

BREEDING BIRD POPULATION DATA SHEET

STUDENT: ____

Locality: ____________________ Dates of study: ________________

Habitat: ____________________ Weather: ________________

Dimensions of grid: ____________ Area of grid: ________________

Date	Hour	Name of bird	Activity-courtship, nesting, feeding, etc.

Remarks: (nests, habitat preference, intraspecific strife, etc.)

Field Work for the Study of Reptile Populations

An area should be selected in which non-poisonous reptiles are known to be common. Abundant cover and a variety of habitats are desirable. Best results may be had during the reproductive seasons when reptiles are much more frequently encountered.

Preferably, a map should be made of the area prior to the study. It is best made using a topographic survey map of the area as a base. On it the plant communities should be noted in detail. Each student is then provided with a copy of the map so that the location of reptiles may be entered on the map as they are observed. The census lines may be entered on the map and distances along them measured. Since an acre is 43, 560 sq. feet, 14 lines 15 feet apart, each extending 209 feet will survey an area approximately one acre in extent. As large an area as time allows should be surveyed.

Form a line, with students 15 feet apart. Move through the area, searching every available habitat. Stones and logs should be overturned, then replaced. When reptiles are found, they should be identified, their measurements and sex recorded, and notes taken as to their habitat. If it is possible to return to the area at a later date (even a year later), the reptiles may be marked, lizards by toe clipping snakes by scale clipping, turtles by notching the carapace. (See p. 73.)

After the area has been covered, all data should be gathered for consideration by the entire class. How many species were found? Are there any clear evidences of habitat preference? Figure out the approximate population per acre of each species, and of all species of reptiles. Do you think you collected all the individuals within your area? What are the average measurements of each species? What is the proportion of juveniles to adults? What is the significance of this?

If a second trip is made to the area, carry out the same collecting procedure, and calculate the population by use of the Lincoln-Petersen index (see p. 74). Would the index be valid if a long period of time intervenes between collections? Individual students may wish to use the class data as a basis for future work in the area.

REFERENCES

Cagle, Fred R. 1953. An outline for the study of a reptile life history. Tulane Stud. Zool. Tulane Univ. 1:31-52.

Schmidt, Karl and Dwight D. Davis. 1941. Fieldbook of snakes. Putnam's Sons, N. Y.

Siebert, H. C. and C. W. Hagen, Jr. 1947. Studies on a population of snakes in Illinois. Copeia 1947(1):6-22.

Smith, Hobart. 1949. Handbook of lizards. Comstock Publ. Co., Ithaca, N. Y.

Stickel, L. F. 1950. Populations and home range relationships of the box turtle, *Terrapene c. carolina* (Linnaeus). Ecol. Monogr. 20:351-378.

Woodbury, A. M., and R. Hardy. 1948. Studies of the desert tortoise, *Gopherus agassizii*. Ecol. Monogr. 18:145-200.

Wright, A. H. and A. A. Wright. 1957. Handbook of snakes of the United States and Canada. 3 vols. Cornell Univ. Press, Ithaca, N. Y.

See ALSO references on p. 150.
For identification manuals, see p. 224.

Field Work for the Study of Amphibian Populations

Any area with large numbers of amphibians is satisfactory for an amphibian population study. If terrestrial salamanders are studied, woods with large numbers of stones or logs may suffice. Ponds or streams may provide suitable habitat for this work if aquatic salamanders or frogs are to be investigated. If possible a map should be made of the area. On it, the plant communities and location of any streams or ponds should be noted. Each student may then be provided with a copy of the map for use while making observations.

Students should move through the area in a broad line, about 10 feet apart, or if a stream is to be studied, each student should be assigned a definite section of the stream. Each amphibian collected should be identified, measured (total length for salamanders, head-vent length for frogs; femur length) weighed (if possible), sexed, and notes made as to coloration or other peculiarities before it is released. Its habitat in which it was found is also noted, and its position marked on the map.

Calculate the population of each species found per acre; of all species per acre. If lines 10 feet apart were used, 22 lines 200 feet long would sample approximately one acre. Compare numbers of adults to juveniles. What is the significance of this proportion of adults to juveniles?

If additional field work is to be carried on later, each am-

REPTILE POPULATION DATA SHEET

STUDENT:______________________ DATE:______________

Location:______________________ Survey Line No.________

Habitat:____

Weather:__________________ Total Area Surveyed____________

No.	Species	Measurements	Sex	Habitat

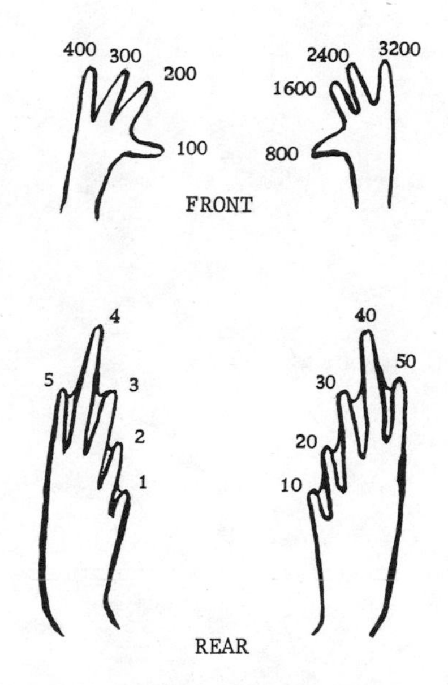

FIGURE 12. METHOD OF MARKING FROGS

(From Martof, Bernard. 1953. Territoriality in the greenfrog, Rana clamitans. Ecology 34:165-175.)

NOTE: Clip at joints to reduce instances of regeneration. Clip off at least two bones. Omit numbers where only one digit would be clipped, such as 10, 20, etc.

FIGURE 13. GRID SYSTEM

	1	2	3	4
A	.	.	.	.
B	.	.	.	.
C	.	.	.	.
D	.	.	.	.

Distances between coordinates should be varied to suit the species with which you are working.

AMPHIBIAN POPULATION DATA SHEET

STUDENT:______________________ DATE:______________

Species:______________________________ Line No.:________

Habitat:____________________ Total area surveyed:________________

Weather:____

No.	Location	Sex	Measurements Snout-vent Femur or total	Weight	Color, etc.

phibian captured should be marked, preferably by toe-clipping (see Fig. 12). The area should be marked in a 5 or 10 meter census grid (see Fig. 13), and for salamanders, individual logs should be marked for future identification. In this way, the total population of each species may be estimated (see p. 74) and relative activity ranges of each species may be indicated. Students may move through the area along a grid line, observing and marking as they go. Each student may be assigned certain numbers to use for marking. The numbers may be repeated with different species.

Those students interested in further work in amphibian populations may make additional surveys of the area to provide more accurate information on population size, activity ranges, and change in weight or size of individuals.

REFERENCES

Cagle, Fred R. 1956. An outline for the study of amphibian life history. Tulane Stud. in Zoo. 4(3).

Highton, Richard. 1956. The life history of the slimy salamander *Plethodon glutinosus*, in Florida. Copeia 1956:75-93.

Martof, Bernard. 1953. Territoriality in the green frog, *Rana clamitans*. Ecology 34:165-174.

Pearson, Paul G. 1955. Population ecology of the spadefoot toad, *Scaphiopus holbrooki* (Harlan). Ecol. Monogr. 25:233-267.

Raney, E. C. 1940. Summer movements of the bullfrog, *Rana catesbeiana* Shaw as determined by jaw-tag method. Am. Midland Naturalist 23:733-745.

See ALSO references on p. 150.
For identification manuals, see p. 224.

Fish Population Study

Because a large number of fishes can be taken easily, a simple population study can be made readily. Select a small area of stream or a small pond where seining can be expected to secure a large proportion of the population. Seine this area thoroughly, placing the captured fishes in a large aerated tank where they will be able to survive temporarily.

When further seining is unproductive, select the species which seems most abundant, or any species which is of special interest. Mark each individual of this species by finclipping or other method (see p. 73 for references on marking fishes) and then release all of the captured individuals. The number of marked fishes must be recorded.

Allow sufficient time for the captured and marked fishes to

return to normal and to reassume their places in the population. This may mean waiting until the next day or next week, or it may be possible to make the first collection early in a class period and the second at the end of the period. The second collection is made in the same manner as the first, but this time marked and unmarked specimens of the species being studied are counted.

Now set up a proportion - Total population (P) : Total marked fishes (M) :: total of the second catch (p) : marked individuals in the second catch (m). Solution of this ratio will give a figure for the total population in the area under study.

Do you think this ratio (known as the Lincoln-Petersen Index) gives accurate results? What refinements could you suggest? What are some possible sources of error in the calculation? What degree of accuracy do you think is required for usefulness in wildlife management or other population studies?

If the sizes of individual fish are recorded, one may gain an idea of the proportion of juveniles to adults. The quantity of each species seined will indicate the relative abundance of the various species in the location seined. By collecting in various habitats, an idea of habitat preferences of each species may be gained. During the reproductive season, color changes may be noted or spawning behavior observed.

REFERENCES

Adams, L. 1951. Confidence limits for the Petersen or Lincoln index used in animal population studies. J. Wildl. Mgmt. 15:13-19.

Cleary, R. E. and J. Greenbank. 1954. An analysis of techniques used in estimating fish populations in streams with particular reference to large non-trout streams. J. Wildl. Mgmt. 18: 461-476.

Fredin, R. A. 1950. Fish population estimates in small ponds using the marking and recovery techniques. Iowa State Col. J. Sci. 24:363-384.

Lagler, Karl F. 1956. Freshwater fishery biology. Wm. C. Brown Co., Dubuque, Iowa.

Rounsefell, George A. and John L. Kask. 1946. How to mark fish. Trans. Am. Fisheries Soc. 73:320-363.

Schumacher, F. and R. W. Eschmeyer. 1943. The estimate of fish population in lakes and ponds. J. Tennessee Acad. Sci. 18:228-249.

See ALSO references on p. 150.
For identification manuals, see pp. 223-224.

Field Work on Invertebrate Populations

Invertebrate populations may be estimated by various methods suitable to the species. A few representative methods may be used in the exercises described below. If a class is large, all these methods may be used at one time. Smaller classes may be able to work only one or two of them in a laboratory period.

During a class field trip, various members may be assigned different sampling duties. One team can be assigned to sweep herbaceous vegetation; another team might sweep shrubs; still another team could sweep trees. A fourth team might sample subterranean arthropods. If it is possible to return to the site several times, the entire class could mark snails for later recapture. For such work, a grid system would probably facilitate a systematic coverage of the chosen area, so that each student could move along a given survey line (see p. 165). The other surveys may also be repeated, but much data may be obtained even from one sampling period. At least 2 hours in the laboratory will probably be required to sort and process arthropods and other invertebrates collected in the field. If return trips are made in less than a week's time, the sweeping and soil samples should not be made in exactly the same location as the previous one. If surveys are made in the same area in the same way and at the same time every year, the data will become still more meaningful.

Sweeping. In areas where herbaceous vegetation dominates, or in grassy openings of woods, insect nets may be used to estimate populations of insects, spiders, and other invertebrates. A net with a 13 inch diameter opening is swept back and forth for 50 strokes, each stroke about 3 feet wide. A step forward is made at the end of each stroke, and the net is quickly turned to prevent the escape of insects within. About half the strokes should be in the upper half of the vegetation, and half in the lower part of the grass. With 50 strokes and 50 paces made in this manner, it is calculated that a sample of arthropods has been made which would equal all the arthropods living in one square meter of vegetation.

In shrubs, 50 strokes are used on all shrub species present. Each species of shrub should be sampled with strokes from near the ground. The more frequent the species of shrub, the more strokes should be made, so that a proportionate part of the 50 strokes are used for each species present.

In trees, outer foliage of each tree species should be swept, again using a proportionate number of the 50 strokes for each tree species. This will of course sample only a small portion of the total tree, since most of it will be out of reach.

All specimens collected may be placed in bags and anesthetized. Specimens should be processed the same day as taken, by mounting on pins or otherwise preserving (see p. 55). Carefully examine both sides of all leaves found in the collecting bags. If sorting cannot be attended to at once, refrigeration will retard loss and deterioration of the specimens.

From data secured by the sweeping technique, one may gain an approximate idea of the total numbers of individuals present; the proportionate numbers of each species; where most insects are located (grass, shrubs, or trees); how populations fluctuate from month to month, year to year, and with weather conditions. Such data become especially meaningful if samples are taken each year in the same location in the same way. For more details and a discussion of the problems of interpreting the results of sampling by sweeping and an analysis of the accuracy of this method, see DeLong (1932) and Shelford (1951).

Mark and recapture. The Lincoln-Petersen index used for vertebrates may also be applied to certain larger invertebrates, especially snails. Snails may be collected and marked by painting with aluminum paint or fingernail polish. Snails are marked and then released. Lines may be set up for each student to follow, so that an area may be systematically examined, and the area covered calculated (see p. 161). Lines 10 feet apart should be satisfactory.

After a week (to allow distribution of marked snails throughout the entire population) the process is repeated, noting the number of snails previously marked, and those found which were unmarked. This is repeated a minimum of 3 times, preferably 6, in order to achieve an accurate estimation of population size. Where repeated collections are made, successive approximations of the total population are obtained, so that the estimates should become closer with each additional sampling. All the marked animals of all previous samplings are included in the M of the formula when successive samplings are made (see p. 74) for a description of the Lincoln-Petersen index). From the results, one may calculate the total population of snails per acre. If different colors are used, it would also be possible to trace movements of individual snails.

Soil sampling. Populations of subterranean invertebrates are most accurately taken by using a steel ring and extracting a core of soil to a given depth with it. (A sample such as 0. 1 sq. meter may be removed with a spade, but it is more difficult to obtain exact size samples.) A can with both ends removed serves as a good tool for this purpose. One having a diameter of 10 cm. (4 inches) such as a tall 1 lb. coffee can, will sample approximately 0.009 sq. meters (0. 01 sq. yard). The can may be depressed to an exact depth, such as 3 inches, and all surface litter and soil within it removed. Surface litter may be placed in a separate container and then put in a

INVERTEBRATE POPULATION DATA SHEET

STUDENT:____________________ DATE:______________

Location:__________________ Community type:____________

Habitat sampled:____________ Air temp:_____ Soil temp:___

Weather:______________ Total area surveyed:____________

Quantity	Taxonomic Group	Percent

Berlese funnel to remove the invertebrates in it (see p. 70). Seives with meshes of various sizes may be helpful in removing invertebrates from the soil (see p. 70). From 3 to 10 such cores should be made. All organisms should be processed as soon as possible after collection. Information similar to that obtained by the sweeping technique may be derived from your data.

REFERENCES

Birch, L. C. and P. D. Clark. 1953. Forest soil as an ecological community with special reference to the fauna. Quart. Rev. Biol. 28:13-36.

Cloudsley-Thompson, J. L. 1958. Spiders, sçorpions, centipedes and mites. Pergamon Press, N. Y.

DeLong, D. M. 1932. Some problems encountered in the estimation of insect populations by the sweeping method. Entomol. Soc. Am. Ann. 25:13-17.

Dowdy, W. W. 1947. An ecological study of the Arthropoda of an oak-hickory forest with reference to stratification. Ecology 28:418-439.

Englemann, Manfred, D. 1961. The role of arthropods in the energetics of an old field community. Ecol. Monogr. 31:221-238.

Foster, Dale. 1937. Productivity of a land snail, Polygyra thyroides (Say). Ecology 18:545-546.

Hairston, N. G. and George W. Byers. 1954. The soil arthropods of a field in southern Michigan: a study in community ecology. Contr. Lab. Vert. Biol. Univ. Mich. 64:1-37.

Jacot, A. P. 1936. Soil structure and soil biology. Ecology 17: 359-378.

Macfadyen, A. 1962. Soil arthropod sampling. Adv. in Ecological Res. Vol. I:1-34.

Shelford, V. E. 1951. Fluctuation of forest animal populations in east central Illinois. Ecol. Monogr. 21:183-214.

Starling, J. H. 1944. Studies of the Pauropoda of the Duke Forest. Ecol. Monogr. 14:291-310.

Wallwork, J. 1959. The distribution and dynamics of some forest soil mites. Ecology 40:557-563.

See ALSO references on p. 150.

For identification manuals, see pp. 221-222.

Section VI
Behavior Studies

Introduction

Of all the subjects of scientific biological inquiry, behavior of animals is among those most readily studied in the field. We may watch live animals in the laboratory and learn a great deal about their behavior, but many times only observation in their natural habitat will tell us how animals really act in nature.

Several facets of behavior are important to the understanding of the life of any animal. These include: the activity patterns of the species - when it is active, and when it rests; its feeding behavior - when it eats, what it eats, and how it selects its food; and reproductive behavior - courtship, breeding, egg laying, family life, etc. Some of these aspects are more easily studied in certain groups of animals than in others, and for many species very little if anything is known about them. The following exercises will indicate the variety of kinds of behavior that may be easily studied and some of the methods employed.

REFERENCES

Bourliere, Francois. 1954. The natural history of mammals. Alfred A. Knopf, N. Y.

Carpenter, C. R. 1958. Territoriality: a review of concepts and problems. In Behavior and Evolution, edited by A. Roe and G. G. Simpson. Yale Univ. Press, New Haven.

Carthy, J. D. 1958. An introduction to the behavior of invertebrates. Macmillan Co., N. Y.

Cloudsley-Thompson, J. L. 1961. Rhythmic activity in animal physiology and behavior. Academic Press, N. Y.

Eibl-Eibesfeldt, I. and S. Kramer. 1958. Ethology, the comparative study of animal behavior. Quart. Rev. Biol. 33:181-211.

Etkin, William, (ed.). 1964. Social behavior and organization among vertebrates. Univ. of Chicago Press. Chicago.

Schiller, Claire H. 1957. Instinctive behavior: the development of a modern concept. International Universities Press, Inc., New York. (Classic papers in the field of ethology.)

Scott, J. P. (ed.) 1950. Methodology and techniques for the study of animal societies. Ann. N. Y. Acad. Sci. 51:1001-1122.

__________. 1958. Animal behavior. Univ. of Chicago Press, Chicago, Ill.

Thorpe, W. H. 1956. Learning and instinct in animals. Methuen & Co., Ltd., London.

Tinbergen, Nicholas. 1951. The study of instinct. Clarendon Press, Oxford.

__________. 1953. Social behavior in animals. Methuen & Co., Ltd., London.

Wynne-Edwards, V. C. 1962. Animal dispersion in relation to social behavior. Hafner Publ. Co., London.

For audio visual aids on behavior, see p. 230.

Squirrel Behavior

On or near most college campuses there are members of one or more species of the family Sciuridae. Gray squirrels, fox squirrels, chipmunks and many species of ground squirrels are diurnal, large enough to be easily observed and often abundant; hence they are excellent subjects for behavioral studies.

Study of the behavior of individuals of one of these species will often tell us much about their role in the community. Such facts as the manner of food selection, eating and storage; times of activity; interrelationships with members of its own and other species; these and other important facts may be learned by observing the animal in its natural habitat. It is also possible to estimate populations of these animals by observation in a limited area.

For best results, observations should be obtained throughout the daylight hours. Each student may therefore be assigned a period during which he is to observe. Routes of travel may be designated, and all squirrels should be noted, their locations charted on a map of the area, and notes made on their activities. The observer must walk slowly and quietly, stopping frequently to look and listen. Even semi-tame park or campus squirrels may "freeze" when disturbed, but if you remain still for a time they will resume their activities.

If continuous coverage of the area is impracticable, each student may be assigned an area and all observations made simultaneously. At least two hours should be spent in the area for adequate population estimates, although shorter periods will give much information about behavior. At the end of the observation period, or in the laboratory following, all observations should be summarized from the maps and data sheets filled out by each student. These notes should include information about habitat preference, feeding and related activities, evidences of mating, intraspecific and interspecific relationships, presence of nests, etc.

SQUIRREL BEHAVIOR DATA SHEET

STUDENT:______________________ DATE:______________

Weather:__________ Time of Tour:_______ Location:________

Kind of Squirrel	Time of Obsv.	Temp.	Wind mph*	Habitat	Activity

Additional notes may be added on the reverse side (interrelationships with other species, nests, etc.).

* Wind: 0-1 mph, smoke rises vertically; 1-3 mph, smoke drifts, but weather vanes do not move; 4-7 mph, leaves rustle; 8-12 mph, leaves and small twigs in constant motion; 13-18 mph, raises dust, small branches move; 19-24 mph, small trees begin to sway; 25-31 mph., large branches in motion, telegraph wires whistle.

From the data secured, we will try to assemble the following information:

1. approximate size of the population of squirrels
2. activity patterns of each species
3. some of the foods consumed or stored by sciurids
4. habitat preferences of each species
5. interrelationships between individuals and with other species
6. miscellaneous behavior of each species

In order to secure more trustworthy data, it would be necessary to repeat these observations at all times of the day, at all seasons and in all kinds of weather as well as in varying habitats over the range of the species. The technique, however, can be demonstrated and practiced as well in one laboratory period as over a longer period.

REFERENCES

Allen, D. L. 1942. Populations and habits of the fox squirrel in Allegan County, Michigan. Am. Midland Naturalist 27:338-379.

Baker, R. H. 1944. An ecological study of tree squirrels in eastern Texas. J. Mammal. 25:8-24.

Flyger, V. F. 1955. Implications of social behavior in gray squirrel management. Trans. 20th N. Am. Wildl. Conf. 381-389.

Gordon, K. 1936. Territorial behavior and social dominance among Sciuridae. J. Mammal. 17:171-172.

Hailman, Jack P. 1960. Notes on the following response and other behavior of young gray squirrels. Am. Midland Naturalist 63:413-417.

Hicks, Ellis A. 1949. Ecological factors affecting the activity of the western fox squirrel, Sciurus niger rufiventer (Geoffroy). Ecol. Monogr. 19:287-302.

Ingles, L. G. 1947. Ecology and life history of the California gray squirrel. Calif. Fish and Game 33:139-158.

Keith, James O. 1965. The Abert squirrel and its dependence on ponderosa pine. Ecology 46:150-163.

Layne, J. N. 1954. The biology of the red squirrel, Tamiasciurus hudsonicus loquax (bangs) in central New York. Ecol. Monogr. 24:227-267.

Moore, J. C. 1957. The natural history of the fox squirrel, Sciurus niger shermani. Bull. Am. Museum Nat. Hist. 113:1-72.

Nichols, John T. 1958. Food habits and behavior of the gray squirrel. J. Mammal. 39:376-380.

Uhlig, H. G. 1956. A theory on leaf nests built by gray squirrels on Seneca State Forest, West Virginia. J. Wildl. Mgmt. 20: 263-266.

See ALSO references on pp. 177-178.
For identification manuals, see pp. 225-226.

Bird Behavior

Birds are good subjects for behavioral studies for several reasons: most of them are diurnal, more or less easily observed, numerous, found in most habitats and are active throughout the year. Furthermore, many species exhibit interesting behavior that stimulates more intensive study.

In a locality where birds are common, demarcate an area of suitable size for the class - one acre per student per hour in the field might be a rough guide. Make a cover map of the area, indicating major vegetational types, and draw grid lines on it. Each student can then be assigned a portion of the area, in which he will walk the grid lines during the period of time allotted. Grid lines can be marked on the land by white flags at 50 to 100 foot intervals.

Although behavior of birds may be observed generally throughout the year, certain types of behavior can be witnessed only at specific periods. For instance migration, mating, and nesting occur for certain limited seasons for each species. If this exercise is held during one of these seasons, it may be modified to take advantage of the opportunity to observe the special behavior. For example, during nesting season, students may be stationed at preselected nesting sites to study the nesting behavior of a particular species. At ordinary seasons, the following procedure may be followed.

Before going into the field, read about various types of bird behavior, such as songs and calls, special display actions, manner of feeding, care of young, nest building, interrelationships with other members of the same species or other bird species, special perching or flying habits. If the members of the class are not well versed in bird identification, one species may be selected for study. Binoculars and a bird identification book should be carried in the field.

Each student will proceed along the assigned route, walking very slowly. If possible, these observations should be made just after sunrise in warm weather, or in early morning in winter. If a bird is observed, stop walking, and slowly make any necessary movements, such as raising binoculars to the eyes. Do not attempt to follow the birds you sight. Mark on your area map each bird observed, indicating by standard signs any activities and the direction in which the bird moved. The signs used in the bird population studies may be used for this study as well. (See p. 158.) Each line will be walked only once during each observation period, but this should be done slowly and with special attention to bird songs, calls and other signs.

In addition to identifying (where possible) the kind of bird, its sex and age, note any behavioral actions exhibited such as:

1. manner of flying- darting, undulating, soaring, flapping, alternate flapping and gliding, etc.
2. special actions- tail wagging, darting from perch and quickly returning, continual hopping about, climbing tree trunks
3. feeding- method of feeding, type of food
4. communication- calls, songs, drumming or pecking - where the birds are while communicating
5. habitat in which seen, and location of the habitat e.g., oak-hickory forest in upper branches
6. distress action - how did it react when it saw you?
7. interrelations with members of its own species
8. interrelations with members of other bird species

In the laboratory, all observations can be gathered and correlated, and the bird population of the area can be estimated. Behavioral patterns of each species should be summarized.

The abundance and activities of any species may vary considerably from one part of the year to another or even from one part of the day to another. For this reason it is advisable to carry out this exercise in the early morning when birds are most active, rather than late in the day. However, observations over long periods of time and under varying conditions are required before the habits of any species can be fully known.

REFERENCES

Allen, Arthur A. 1961. A book of bird life. D. Van Nostrand Co., N. Y.

Bellrose, Frank C. 1963. Orientation behavior of four species of waterfowl. Auk 80:257-289.

Emlen, J. T. 1950. Techniques for observing bird behavior under natural conditions. Ann. N. Y. Acad. Sci. 51:1103-1112.

Griffin, Donald R. 1964. Bird migration. The biology and physics of orientation behavior. Science Study Series S 37. Doubleday & Co., Inc., N. Y.

Howard, E. 1948. Territory in bird life. Collins, London.

Hutson, H. P. 1956. The ornitholgist's guide. Philosophical Library, N. Y.

Moynihan, M. 1956, 1958, 1959. Notes on the behavior of some North American gulls. I, II, III, IV. Behavior 10:126-179; 12:95-182; 13:112-130; 14:214-239.

Nice, M. M. 1941. The role of territory in bird life. Am. Midland Naturalist 26:441-487.

Orians, Gordon H. and Mary F. Wilson. 1964. Interspecific territories of birds. Ecology 45:736-745.

Noble, G. K. 1939. The role of dominance in the social life of birds. Auk 56:263-273.

See ALSO references on pp. 177-178.
For identification manuals, see pp. 224-225.

Amphibian Breeding Habits

Spring ponds often contain a number of species of amphibians from early spring to midsummer. A study of these ponds can be a continuous project through these months, or can be a single field trip at the time when the greatest number and variety of amphibians will be present.

If possible, at least one night field trip should be planned, since frogs and toads will be calling and courtship procedures can be observed. A strong flashlight will blind amphibians, causing them to remain motionless as you approach for closer observation. Salamanders are abroad at night, especially just after or during the first warm rainy night of the spring. Frequently the ice may still be disappearing from the ponds when the amphibian breeding commences.

Record the temperature of the air and water, weather and relative humidity. What species are actually found in the pond? Look for others among the vegetation, under logs and stones around the pond edges. Determine what species deposited the various egg masses. Male salamanders deposit gelatinous structures called spermatophores (sperm-bearers). These bear a cap of sperm-filled material at the tip, and at the end of courtship the female picks up the cap with the lips of the cloaca, thus ensuring internal fertilization. Look for spermatophores on leaves at the bottom of the pond. Try to find individuals in amplexus, or performing courtship. What differences do you observe between the types of egg masses found, including numbers of eggs per group, and position in the water? Can you observe differences in the sexes of salamanders? Of frogs and toads? Are these amphibians permanent residents of the pond? If not, where do they spend the rest of the year?

A technique for collecting salamanders during the reproductive season is to erect a low (3 or 4 inch) fence around a good part of the perimeter of a pond in which salamanders are known to breed. At intervals the fence should funnel into a trap, such as a sunken can. The fence could be constructed out of any suitable material, but 1/4 inch hardware cloth will do. Since most movement toward the pond will occur during the night, a daily morning survey of the traps is all that is needed. Such traps would facilitate a longer term study (class or individual project) of the breeding times of local salamanders.

AMPHIBIAN BEHAVIOR
DATA SHEET

STUDENT:____________________ DATE:________________

Location:____

Time:__________ Air Temp.:_________ Water Temp.:_________

Weather:____

Species collected	Sex	Spermatophores		Eggs	
		No.	Position	No.	Position

Remarks: (description of spermatophores, eggs; species calling; courtship behavior, etc.)

REFERENCES

Bishop, Sherman C. 1943. Handbook of salamanders. Comstock Publ. Co., Ithaca, N. Y.

Blair, Frank W. 1958. Mating call in the speciation of anuran amphibians. Am. Naturalist 92:27-51.

Davis, William C. and Victor C. Twitty. 1964. Courtship behavior and reproductive isolation in the species of _Taricha_ (Amphibia, Caudata). Copeia 1964:601-609.

Martof, Bernard S. 1953. Territoriality in the green frog, _Rana clamitans_. Ecology 34:165-174.

Martof, Bernard S. and Eric F. Thompson, Jr. 1958. Reproductive behavior of the chorus frog, _Pseudacris nigrita_. Behavior 13:243-258.

Smith, R. E. 1941. Mating behavior in _Triturus torosus_ and related newts. Copeia 1941: pp. 255-262.

Storm, R. M. and R. A. Pimentel. 1954. A method for studying amphibian breeding populations. Herpetologica 10:161-166.

Wright, A. H. and A. A. Wright. 1949. Handbook of frogs and toads. Comstock Publ. Co., Ithaca, N. Y.

See ALSO references on pp. 177-178.
For identification manuals, see p. 224.
For identification record, see p. 230.

Nocturnal Activity

An important behavioral pattern of any animal species is its time of activity. Many animals are active at night, yet all animals that are nocturnal are not active at the same hours. Some are active early in the evening, others late in the evening, while some appear in the early morning hours. An appreciation of the complexity of the interrelationships of animals can be gained by a night's study of the activity of animals in a wood lot.

A trail will be set up by the instructor through a woods containing several stumps or logs. Such a woods would preferably be a climax community, where leaf litter is thick, and animals abundant. Commencing at dusk, the class will cover the trail, stopping at each log or stump to note the kinds and numbers of animals found and their positions on trunks and stumps. Each invertebrate may be marked with a dab of aluminum paint. The trail will be retraced at regular periods throughout the night. As the night wears on, it will be seen that various species appear at certain times and then move up and finally down the trees. Light traps (see p. 55) and sugar lures may also be used and checked at regular intervals to gain an idea of the time of activity of various species. Sugar lures may be prepared with brown sugar and/or molasses, mixed with bananas or crushed apples. The mixture is allowed to age a bit, until slightly fermented, when it will produce an attractive odor. It is then brushed on suitable trees along a woodland trail.

When not engaged in observing animals on the trail, the students will make hourly records of wind velocity, relative humidity, evaporation rate and temperature.

From the data gathered, you should be able to determine:

1. the period during which the largest numbers of animals are active
2. the period during which the most species are active
3. any correlation of activity of animals with the time, temperature, humidity, or other physical factors

What advantage might it be for a species to have a genetically determined hour of activity (modifiable by local environmental conditions)?

REFERENCES

Calhoun, John. 1944, 1945, 1946. Twenty-four hour periodicities in the animal kingdom. J. Tennessee Acad. Sci. 19:179-200; 19: 252-262; 20:228-332; 20:291-308; 20:373-378; 21:208-216; 21:281-282.

Cloudsley-Thompson, J. L. 1961. Rhythmic activity in animal physiology and behavior. Academic Press, N. Y.

Griffin, D. R. 1958. Listening in the dark. Yale Univ. Press, New Haven.

Milne, Lorus J. and Margery Milne. 1956. The world of night. Harper Bros., N. Y.

Park, Orlando. 1935. Studies in nocturnal ecology III. Recording apparatus and further analysis of activity rhythm. Ecology 16:152-163.

__________, 1940. Nocturnalism. The development of a problem. Ecol. Monogr. 10:486-536.

__________, J. A. Lockett and D. J. Myers. 1931. Studies in nocturnal ecology I. Studies in nocturnal ecology with special reference to climax forest. Ecology 12:709-727.

__________, and J. G. Keeler. 1932. Studies in nocturnal ecology II. Preliminary analysis of activity rhythm in nocturnal forest insects. Ecology 13:335-347.

__________, and H. F. Stronecker. 1936. An experiment in conducting field classes at night. Ohio J. Sci. 36:46-54.

Williams, C. B. 1940. An analysis of four year's captures of insects in a light trap. II. The effect of weather conditions on insect activity; and the estimation and forecasting of changes in the insect population. Trans. Roy. Entom. Soc. London 90:227-306.

See ALSO references on pp. 177-178.
For identification manuals, see p. 224.

NOCTURNAL ACTIVITY
DATA SHEET

STUDENT:______________________________ DATE:____________________

Location:________________________Weather:_____________________________

TOUR I			
Time ____________ Air temp., surface ________ Rel. hum., surface ____________ Air temp., 3 ft. ________ Rel. hum., 3 ft. ____________ Wind velocity ________			
Species	Quant.	Habitat	Distance from ground

NOCTURNAL ACTIVITY
DATA SHEET

STUDENT:____________________ DATE:________________

Location:____________________ Weather:________________

TOUR II			
Time ________ Air temp., surface ________ Rel. hum., surface ________ Air temp., 3 ft. ________ Rel. hum., 3 ft. ________ Wind velocity ________			
Species	Quant.	Habitat	Distance from ground

NOCTURNAL ACTIVITY DATA SHEET

STUDENT:____________________ DATE:____________

Location:________________ Weather:________________

TOUR III

Time	________	Air temp., surface	________
Rel. hum., surface	________	Air temp., 3 ft.	________
Rel. hum., 3 ft.	________	Wind velocity	________

Species	Quant.	Habitat	Distance from ground

NOCTURNAL ACTIVITY
DATA SHEET

STUDENT:______________________ DATE:______________

Location:_________________ Weather:__________________

TOUR IV			
Time ________ Air temp., surface ________ Rel. hum., surface ________ Air temp., 3 ft. ________ Rel. hum., 3 ft. ________ Wind velocity ________			
Species	Quant.	Habitat	Distance from ground

NOCTURNAL ACTIVITY
DATA SHEET

STUDENT:________________________ DATE:________________

Location:____________________Weather:____________________

TOUR V

Time	________	Air temp., surface	________
Rel. hum., surface	________	Air temp., 3 ft.	________
Rel. hum., 3 ft.	________	Wind velocity	________

Species	Quant.	Habitat	Distance from ground

NOCTURNAL ACTIVITY
DATA SHEET

STUDENT:_____________________________ DATE:____________________

Location:__________________________Weather:__________________________

TOUR VI

Time ____________ Air temp., surface ____________
Rel. hum., surface ____________ Air temp., 3 ft. ____________
Rel. hum., 3 ft. ____________ Wind velocity ____________

Species	Quant.	Habitat	Distance from ground

Behavior and Distribution of Invertebrates

The behavior of animals is one of the factors affecting their local distribution. Their distribution may therefore reflect their reaction to various conditions of the environment, such as moisture, light or temperature. On the other hand, their distribution may be affected by their reactions to other individuals of the same species or of a different species.

Organisms may be distributed in one of 3 general ways: evenly, randomly, or unevenly spaced. If a species is evenly spaced, the distance from one individual to another would be more or less constant. A randomly spaced population would have some individuals close to another in some spots, farther in another. Finally, an unevenly spaced population would show tendencies towards aggregation, or clumping. (See Odum, 1959, for a discussion of the possible survival values of each of these types of distribution.) Evenly and randomly spaced populations are not too common in nature, while unevenly spaced populations are the usual type. The exercise here will probably illustrate the latter two types of distribution.

If small boards are placed in a grid (see p. 165) with the interval between boards about 100 feet, invertebrates will seek shelter under the boards. In one study (Cole, 1946), spiders, diplopods, chilopods, isopods, and beetles were found under boards placed in a forest. The size of the boards might be from 4" x 4" (such as mammal scat boards, described on p. 155) up to 12" x 12". The boards should be placed in as uniform an environment as possible. After placing at least 100 boards in a grid, a map is made, indicating cover type, slope or other significant physical condition. The boards should be inspected at regular intervals, and the number of each species of invertebrates located under each board recorded. (It is important not to lump the information from one board with that of another board.) If time allowed, it would be desirable to take the temperature of the soil under each board, and make an estimate of the moisture content and soil texture under each board. The class members could each take a proportionate number of boards to record data on, thus reducing the time necessary. At the end of the observation period, all data are collected and summarized.

A table should be set up for each species observed, listing the number of individuals under a board, and the number of boards having that number of individuals under it. If a species is randomly distributed, you would expect many boards with no individuals, a few boards with one individual, fewer boards with 2 individuals, still fewer boards with 3 individuals, etc. The larger the population, the higher the totals in each category will be, of course. Clumped distribution would be shown by the tendency to have larger numbers of boards with more than one individual under them compared to random distribution.

If certain conditions are met, including large numbers of boards (so that the chance a particular individual will occur under a given board is small) and a relatively high population, then a certain calculation will indicate whether or not the distribution of a population is random. Thus, where

X = the number of individuals of one species under any one board

$\bar{X}$ = the average number of individuals of one species under all boards

n = the number of boards

then if the sum of $(X - \bar{X})^2$ divided by $\bar{X}(n - 1)$ is greater than $2\sqrt{\frac{2n}{(n-1)^2}}$ at the 5% level of probability we may consider that the population is not randomly distributed. (It may in fact, be more or less evenly distributed.) For a derivation of this formula, and examples of its use on arthropod distribution, see Andrewartha and Birch, 1954. The formula is a simplification of a test for randomness, called the Poisson series*. The analysis of randomness of distribution

* For those who have a knowledge of statistics, the expected values for the frequency of boards having a given number of individuals under them may be calculated from the formula for the Poisson distribution:

$$P_r = \frac{s^r e^{-s}}{r!}$$

where s = mean number of individuals per board
r = actual number of individuals per board
P_r = probability of having r individuals in a given sample (and therefore the expected frequency of samples containing r individuals)
e = Naperian base of logarithms
$r!$ = r factorial

Thus, from the Poisson series,

the frequency of boards with 0 individuals (r=0), $P_0 = e^{-s}$
" " " " " 1 " (r=1), $P_1 = (e^{-s})s$
" " " " " 2 " (r=2), $P_2 = (e^{-s})\frac{s^2}{2}$
" " " " " 3 " (r=3), $P_3 = (e^{-s})\frac{s^3}{6}$
" " " " " 4 " (r=4), $P_4 = (e^{-s})\frac{s^4}{24}$
" " " " " etc. " etc. etc. etc.

must be used with caution, since under certain circumstances it will not apply. Data collected on different days or from different areas should be analyzed separately.

In Cole's study, only spiders appeared to be randomly distributed; sow bugs, centipeds, millipedes and beetles were aggregated. If you find there is non-random distribution, the next step is to determine the causes. It might be due to the physical requirements of the species which are met only under certain boards, or it might be due to reproductive or other behavioral traits not directly connected with the physical environment. It might be other members of the species which attract more individuals of the species. A careful analysis of your data might suggest possible causes for the types of distribution you find.

REFERENCES

Allee, W. C. 1926. Studies in animal aggregations. Causes and effects of bunching in land isopods. J. Exp. Zool. 45:255-277.

__________, A. E. Emerson, O. Park, T. Park, and K. P. Schmidt. 1949. Principles of animal ecology. W. B. Saunders Co., Phila.

Andrewartha, H. G. 1961. Introduction to the study of animal populations. Univ. of Chicago Press, Chicago.

__________ and L. C. Birch. 1954. The distribution and abundance of animals. Univ. of Chicago Press, Chicago, Ill.

Auerbach, Stanley I. 1949. A preliminary study on certain deciduous forest centipedes. Am. Midland Naturalist 42:220-227.

Cole, Lamont C. 1946. A study of the cryptozoa of an Illinois woodland. Ecol. Monogr. 16:49-86.

Odum, Eugene. 1959. Fundamentals of ecology. W. B. Saunders Co., Phila., Pa.

Salt, G. and F. S. J. Hollick. 1946. Studies of wireworm populations. II. Spatial distribution. J. Exp. Biol. 23:1-46.

Snedecor, George W. 1956. Statistical methods. 5th ed. Iowa State Col. Press, Ames, Iowa.

The values calculated for frequency of samples having 1, 2, 3, ... number of individuals are the expected values if the distribution is random. These are used to run a chi square test using observed values from the boards, and using n-2 degrees of freedom (n-2 rather than n-1 because expected frequencies are calculated from totals and means). Here, n = total number of boards.

Behavior of An Orb-weaving Spider

In most parts of the United States, orb-weaving spiders are easily found throughout the warmer parts of the year. One of the common species is the golden orb-weaver, Argiope aurantia, which in some areas is extremely abundant. This exercise is best performed where at least several webs may be observed.

In a suitable habitat chosen by the instructor, divide into small groups and locate webs of an orb-weaver. Observe and record for each web the following data:

1. position of the spider in the web
2. orientation of the web in respect to direction
3. orientation of the web in relation to the earth, e. g., vertical, horizontal, or at what angle
4. draw as accurately as possible a diagram of the web.

Gather a number of insects of suitable size (e.g., small grasshoppers, leafhoppers, flies, etc.). Without casting a shadow on the web or disturbing it in any way, toss an insect so that it is caught in the web. Observe and record the activities of the spider. How long does it take to reach the prey? Does it approach directly or by a devious route? What is its first activity in respect to the prey? Does it bite at once? How does it immobilize the prey? Does it then begin to feed immediately or does it leave its prey? Repeat the operation as many times as required to observe and understand fully the responses of the spider.

Cut or break one of the main supporting strands of the web. Observe and record the response of the spider. Does it immediately set about repairing the break? How does it operate? Where does the silk for repair come from? Is the repaired web identical in form and structure to the original one? What happens if you break more than one thread simultaneously?

By carefully placing a wire hoop under the web, you can fasten the main supporting threads to the hoop with cellulose tape or masking tape. Capture the spider and place it in a jar, and then collect the net with as little damage as possible. You can then transport both spider and web to the laboratory. Here you may make as many observations on its behavior as time and ingenuity permit. What happens if you alter the orientation of the web with respect to direction? Light? Plane of the earth? Will the spider

* Parts of this exercise have been used by the senior author in field biology courses. Some of the techniques have been suggested by an exercise in Cummins et al., 1965.

feed normally with the position of the web altered? What effect do changes in temperature, humidity, chemical stimuli, etc., have on the spider's behavior?

A spider removed from its web and placed in a suitable artificial environment will build a new web. If time permits, observe and record the exact order of events in web-building. Compare your observations with those given in one of the references below. Will the spider build in the dark? If so, does this web look the same as one built in the light?

REFERENCES

Carthy, J. D. 1958. An introduction to the behavior of invertebrates. G. Allen, London.

Cummins, K. W., L. D. Miller, N. A. Smith and R. M. Fox. 1965. Experimental entomology. Reinhold Publ. Corp., N. Y.

Curtis, Helena. 1965. Spirals, spiders and spinnerets. Amer. Scientist 53:52-58.

Lougee, L. B. 1963. The web of the spider. Cranbrook Inst. of Science, Bulletin 46.

Witt, P. N. 1957. Spider webs and drugs. Scientific American Books, Set No. 1: Animals. Simon and Schuster, N. Y.

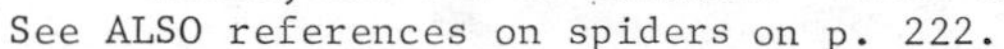

See ALSO references on spiders on p. 222.

Section VII
Projects for Field Study

General Procedure

As a part of the work of this course, your instructor may wish to assign special field projects. These field projects are designed to accomplish several things:

1. to give you practice in field work and techniques
2. to give you experience with some organisms of your area
3. to give you more intimate knowledge of one area of the literature of biology

The projects are not necessarily designed to require original work, so that a literature survey is not a prerequisite first step in your project. However, you will want to consult references to aid in formulating your approach to your project and to help you throughout the course of your study. (See Benton and Werner, 1965, Chap. XII.) Note the references listed at the ends of the sections and after the laboratory exercises. They are representative papers and may give you ideas as well as additional references. If your school does not have the journals in which the articles appear, they may usually be obtained through inter-library loan. The articles referred to may be photostated if they are needed longer than the period allowed by the loan privilege. You can find which libraries have a particular journal in the Union List of Serials.

CLASS PROJECT

Projects may be approached in one of three general ways: as a class project, as a team endeavor, or as an individual study. A class project has certain advantages. It allows the accumulation of data on one topic, and thus may be able to produce more significant results. The amount of work required of each student may be easily limited, making the assignment of a project more feasible where time is short. It tends to define the project for the students, which may make supervision of the work by the instructor simplcr. It is also easy to assign various members of the class to work in teams. Finally, over a period of years, such projects may be continued so that each succeeding class may profit by the information gained in previous years.

A group project should be outlined to the whole class. Various projects within the limits of the class project can be listed, and each student may select one which suits him. For example, a class project might be a study of insect populations (see p. 171) Individuals or teams may then take different aspects of the problem, such as physical measurements of the environment, collection and preservation of insects from one vegetational type, etc. Students might also suggest new facets of the whole problem. Data collected by each individual or group may be duplicated and distributed to the entire class for interpretation.

TEAM PROJECT

Team projects have the advantage of allowing students to engage in problems which one student by himself could not physically handle, such as one calling for the operation of a 10 foot seine. Also, several people as a team may choose a slightly larger problem and accomplish it within a limited time period. Teams, like individuals, must decide on their projects early in the semester, and should select a topic suitable to their interests, abilities and time available. See the comments below pertaining to conduct of a problem by individuals for further suggestions.

INDIVIDUAL PROJECT

An individual project also has certain merits. It allows a student the widest possible choice of topics. It gives him experience in setting up and carrying through a project on his own. It is especially suitable where there is more time for the project to be accomplished.

The student should consult with the instructor on choice of a problem. It is essential that the project be small and capable of being completed within the time available. The problem can often be narrowed down by restricting the taxonomic group of the study, or by restricting the geographical area under consideration. Start the project at the beginning of the semester - do not think of this as a "term paper". A project rarely can be done satisfactorily as a concentrated effort within one or two weeks. Choose a specific problem and decide on specific methods for its solution. In this connection, see Benton and Werner, 1965, Chap. XIII. The instructor will help you with the necessary equipment, as well as help on specific procedures and techniques.

CONCLUSION OF THE PROJECT

A final report on your project will be expected at the end of the course. Before you write it up, your instructor may wish you to hand in a rough draft or outline so that he may check your method of presentation. The report should follow the organization and style of a scientific publication such as Ecology, which will be available for you to study. Note particularly the following:

1. Order of presentation of findings
   a. introduction and statement of the problem
   b. mention of significant literature
   c. description of methods and materials used
   d. presentation of findings, data
   e. conclusion or discussion
   f. summary or abstract
2. References
   a. the manner of citing literature
   b. footnotes are not used
   c. use "References" rather than "Bibliography"
3. Content is more important than length. Do not pad your results or use unnecessary words.
4. Include negative as well as positive material and note any special problems you had or things that went wrong. This is helpful to anyone who may later be interested in the same or a similar problem.

REFERENCES

Benton, Allen H., and William E. Werner, Jr. 1965. Field biology and ecology. McGraw-Hill Book Co., N. Y.

Committee on Form and Style of the Conference of Biological Editors. 1960. Style manual for biological journals. Amer. Inst. of Biol. Sci., Washington, D. C.

Trelease, Sam F. 1958. How to write scientific and technical papers. The Williams and Wilkins Co., Baltimore, Md.

Suggestions for Projects

TAXONOMY

Collect and identify all the plants or animals of a taxonomic group in a given area. The collection should be supported by an annotated list together with a bibliography of books on identification of the group, especially for the local area.

Analyze the identifying features of a restricted group of organisms to determine variability of these features and the suitability of their use in keys. The analysis should include a review of the characteristics used in existing keys to these organisms, and where possible suggestions for improvement of the keys should be made. See pp. 10-14.

TERRESTRIAL COMMUNITIES AND SUCCESSION

Conduct a thorough survey of an area undergoing succession. Restrict your survey to one taxonomic group, which should be collected, preserved, and identified. Collections should be made over a period of time to include specimens present at various seasons. See pp. 78-92.

Determine the rate of growth of woody plants of a given species in various stages of succession. See pp. 81-87.

Study the microsuccession in rotten logs, galls, cow dung, or other niches. See pp. 92-94.

Determine what plants serve as pioneers in your locality. Which ones can colonize soil with organic material? Rock? Is there a variation of pioneer species in differing types of pre-existing soil? See pp. 78-80.

Sample a climax forest in different areas, or in slightly different physical situations, and compare the species composition. See pp. 85-87.

Compare the animal and/or plant species along a transect passing from one type of a community to another. See pp. 70-71

Make a comparative study of animal species (such as all birds) present in an old field with clumps of shrubs interspersed among the herbaceous vegetation, and a nearby forest. Determine if an "edge effect" is present. See Odum, 1959. Fundamentals of ecology.

Examine all the woody species in a field undergoing secondary succession. Which are seedlings, which are sprouts? Can you determine the ages of individuals? If so, which, in general are older, trees or shrubs? See pp. 81-85.

AQUATIC COMMUNITIES AND SUCCESSION

Study the diurnal vertical migrations of phytoplankton organisms, or larger animals.

Compare the productivity of two ponds or lakes; or study the change in productivity of one pond or lake over a period of time. See pp. 135-138.

Make a study of the aufwuchs or periphyton (organisms living on and about aquatic plants). The study could compare the organisms found at different levels in the water, on different species, or in different aquatic communities.

Compare a particular taxonomic group of organisms in swift streams and slow streams or in lakes and ponds, etc. Note which species present are characteristic for each community, which are ubiquitous. Note anatomical features allowing adaptation to their environment. See pp. 116-127.

Examine the physical conditions (oxygen, carbon dioxide tension, substrate type, pH, turbidity, etc.) in the zones of a pond or lake and correlate them with the kinds of organisms found in each zone. See pp. 116-121.

Study one physical factor such as the current of a stream or the depth of a temporary pond through a season and correlate physical changes with changes in the population of one group of organisms. See pp. 116-127.

POPULATIONS

Collect one species of fish, and by reading the scales, determine growth rate and correlate this with abundance of the fish in each age class. See pp. 169-170 and Trautman, Ohio, Fishes (for growth data) and Lagler, Freshwater Fishery Biology (for technique).

Make an analysis of the habitats of 2 closely related species. What are the differences in requirements of each? The similarities? See pp. 70-72, 201-203.

Determine the age and sex composition of a species of animal in a restricted habitat. Make comparisons with another locality or the same locality in another season. See pp. 162-169.

Determine the daily movements of an amphibian, reptile, or mammal using conventional or an original technique. See pp. 150-156; 160-169.

Compare methods of sampling plants, such as transect, quadrat, arms-rectangle method, etc. See pp. 70-71.

Compare dropping board, snap trap and live trap methods of sampling small mammal populations. See pp. 150-156.

Determine the fluctuation of one species of phytoplankton or zooplankton in a pond, lake, or stream throughout a season. See pp. 51-52.

Determine the population variation of one kind of organism in different aquatic habitats, such as stream and pond, or in a wave-washed area compared to a protected area. See pp. 169-170.

Compare the soil animals from 2 dissimilar communities, or from the same community from different soil levels, or from the same level at different times or seasons of the year. See pp. 72, 172-173.

BEHAVIOR

Study the food selection of a species. Analyze stomachs of a predator (such as frogs). Collect the predator and if possible its food at different times of the day, or at different seasons to determine variation. Does the animal select food, or eat food most easily caught? See pp. 62-66.

Investigate the feeding behavior of a species. How is food secured; what distance is traveled to obtain it; what are its relationships with other members of its species and other species while feeding? See pp. 178-183.

What are the sex relationships of a species? Observe mating behavior, courtship, care of young, reaction to other members of the species and other species during the reproductive season. See pp. 182-183.

Determine the activity periodicity of an animal. What time of day is the animal active, when does it rest? Is its activity rhythm innate or influenced by its environment? See pp. 178-181; 187-188.

Study the home range of a species. What part of its home range, if any is defended? See pp. 73; 150; 155; 157; 169; 172.

PHYSIOLOGICAL ECOLOGY

How does the slope of an area affect the presence of a plant or animal species, or the species composition of a community? What physical factors may be involved? (Such as light, moisture, pH, soil type, etc.) See pp. 77-92.

Determine the variation in oxygen tension (or other physical condition) in different bodies of water representing different community types, at different water levels, or at different times of the day or season, or at varying distances from shore. What would the significance of your findings be on organisms in these areas? See pp. 116-127; 135-138.

What effect does relative humidity have on growth of leaves of a particular species; on the type of insect in a cave, log, etc.? See pp. 92-94; 187-188.

Investigate the correlation of soil conditions (pH, moisture, soil profile) with stages of succession or climax types. See pp. 77-92.

What are the light requirements of a species of plant? Can it tolerate shade or full sunlight as a seedling? As a mature plant? See pp. 78-87.

What light intensity regulates the activity of a particular species of animal? When does it become active? If nocturnal, is its activity influenced by the amount of moonlight? See pp. 187-188.

Compare the physical aspects of a burned-over forest site with those of an unburned site with respect to soil and air temperatures, pH of the soil, soil profile, etc. What plant species survived the fire and are now surviving the new environmental conditions? Are any new species invading? Which species sprouted, which did not?

Section VIII

Selected Biological Literature

The literature of biology has become so vast that it is totally impossible to present any complete selection. Within the limits of space, we have had three aims: 1) to introduce the student to some of the more important bibliographic aids, which may serve as keys to the rest of the literature; 2) to list a few of the journals of ecology, natural history, and field biology, from a variety of disciplines and a number of different countries; and 3)to list some of the more useful identification manuals, field guides, general references, and natural history books of interest to the field biologist.

We suggest that the instructor devise library exercises which will insure student use of as many of these publications as possible. For students who look forward to a career in biology, an early and thorough grasp of the literature is an essential. For those who plan to teach, a knowledge of the less technical literature is necessary, particularly of those works which will aid in identification of organisms. Biological books and journals are of ever-growing importance, and the student should be given ample opportunity to develop the knowledge and techniques necessary for effective use of these resources.

Bibliographic Aids for the Field Biologist

Bay, J. C. 1910. Bibliographies of botany. Progressus Rei Boanicae 3:331-456.

Bibliographic Index. 1945 and following. H. W. Wilson, N. Y. (An index to published bibliographies: supplements are published every few years).

Bibliography of Agriculture. 1942.and following. U. S. Dept. of Agriculture, Washington. (Monthly.)

Biological Abstracts. 1926 and following. Philadelphia, Pa. (Now published biweekly).

Botanical Abstracts. 1918-1926. Baltimore Md.

Dean, Bashford. 1916-1923. Bibliography of Fishes. Amer. Mus. of Nat. Hist., N. Y.

Index to Publications of the U. S. Dept. of Agriculture. 1943 and following. Washington, D. C.
Index to the literature of economic entomology. 1890 and following. Washington, D. C.
Jackson, B. D. 1881. Guide to the literature of botany. Longmans, Green and Co., Ltd., London.
Lawrence, G. H. M. 1951. Taxonomy of vascular plants. The Macmillan Co., N. Y.
Publishers Trade List Annual. 1873 and following. R. R. Bowker Co., N. Y. (Includes volumes giving listings by publisher, by author, by title and by subject).
Rehder, A. 1911-1918. The Bradley bibliography: a guide to the literature of the woody plants of the world published before the beginning of the twentieth century. 5 vols. Harvard Univ. Press, Cambridge, Mass.
Smith, Roger C. 1962. Guide to the literature of the zoological sciences. 6th ed. Burgess Publ. Co., Minneapolis, Minn.
Strong, Reuben M. 1939-1946. A bibliography of birds. 3 Vols. Field Mus. of Nat. Hist., Chicago, Ill.
United States book catalog and the cumulative book index. 1928 and following. H. W. Wilson Co., N. Y.
Wildlife Abstracts, 1935-1951. U. S. Fish and Wildlife Service, Washington; edited by Neil Hotchkiss.
Wildlife Abstracts, 1952-55. U. S. Fish and Wildlife Service, Washington; edited by Lucille Stickel.
Wildlife Abstracts, 1956-60. U. S. Fish and Wildlife Service, Washington; edited by Nicholas J. Chura.
Wildlife Review. 1935 and following. U. S. Fish and Wildlife Service, Washington.
Wood, Casey. 1931. An introduction to the literature of vertebrate zoology. Oxford Univ. Press, London, England.
Zoological Record. 1864 and following. Zoological Society of London, London, England.

Biological Journals for the Field Biologist

A complete list of journals of interest to the field biologist and ecologist would certainly include hundreds, perhaps thousands, of entries. Rather than ignore the subject entirely, we have chosen to list a few of the more important journals, both American and foreign, in the hope that these lists will provide a point of departure for the student and lead him to a further search. Probably one of the best "second steps" is a study of the list of periodicals reviewed in Biological Abstracts. Such a list is published each year, and it includes most, if not all, of the important biological journals in the world. As a first step, acquaintance with the journals listed here will provide the student with a glimpse of the vast world of ecological literature, and introduce him to many of the great scientists and significant publications with which he will need to become increasingly familiar.

Advances in Ecological Research. 1962-
American Fern Journal. 1910-
American Journal of Botany. 1914-
American Midland Naturalist. 1909-
Annals of the Entomological Society of America. 1908-
The Auk. 1884- (Ornithology)
The Condor. 1899- (Western Ornithology)
Copeia. 1913- (Reptiles, Amphibia and Fishes)
Ecology. 1920-
Ecological Monographs. 1930-
Evolution. 1947-
Herpetologica. 1945- (Reptiles and Amphibia)
Journal of Mammalogy. 1919-
Journal of Parasitology. 1914-
Psyche. 1874- (Entomology)
Science. 1883-
Systematic Zoology. 1952-
Transactions of the American Society of Limnology and Oceanography.
Wildlife Monographs. 1958-
Wilson Bulletin. 1889- (Ornithology)

JOURNALS OF OTHER COUNTRIES

Acta Hydrobiologica. Poland.
Acta Societas pro Fauna and Flora. Finland.
African Wildlife. South Africa.
Anales del Instituto de Biologia. Mexico.
Annals and Magazine of Natural History. England.
Animal Behaviour. England.
Archiv fur Hydrobiologie. Germany.
Australian Journal of Marine and Freshwater Resources.
Biota. Peru.
Byulleten Moskvoskogo Obshchestva Ispytatelei Prirody otdel Biologicheskii. Russia. (Bulletin of the Moscow Society for the Investigation of Nature, Biological Section).
Ecological Review. Japan.
Journal of Animal Ecology. England.
Journal of Ecology. England.
Nature. England (similar to the American journal, "Science").
Oikos. Denmark. (ecology).
Sbornik Rabot po Ikhtiologii I Gidrobiologii. Russia.
Suomen Riista. Finland. (wildlife).
Zeitschrift fur Saugetierkunde. Germany. (mammals).
Zoologicheskii Zhurnal. Russia.

While the following lists are rather extensive, they are not intended to be either complete or definitive. In particular, we have omitted a large number of excellent state and regional works in a variety of different fields. Almost every state has a bird book, many have mammal books, and there are numerous state books on a variety of animal and plant groups, particularly those of economic importance. These books are of primary importance to the biologist, but there are so many that it would be completely impractical to attempt a complete list here. The student should take pains to familiarize himself with books and serial publications relating to his particular area, and to secure as many of them as are available in fields of his interest.

The few state books that we have included are there because we feel that they are of special value even in areas outside the states they cover, or because the regional works do not cover the states themselves (e.g., books on Alaska and Hawaii are included because books covering the United States are most frequently restricted to the 48 contiguous states).

General References and Manuals for Identification

ALGAE AND FUNGI

Bessey, Ernst A. 1950. Morphology and taxonomy of fungi. McGraw-Hill Book Co., N. Y.

Chapman, Valentine J. 1961. Algae. St. Martin's Press, Inc., N. Y.

Christensen, Clyde M. 1955. Common fleshy fungi. Burgess Publ. Co., Minneapolis.

Dawson, E. Yale. 1956. How to know the seaweeds. W. C. Brown Co., Dubuque.

Duncan, U. K. 1959. Guide to the study of lichens. Scholar's Library, N. Y.

Fergus, Charles L. 1960. Illustrated genera of wood decay fungi. Burgess Publ. Co., Minneapolis.

Fink, Bruce. 1960. The lichen flora of the United States. Univ. of Michigan Press, Ann Arbor.

Fischer, G. W. 1953. Manual of North American smut fungi. Ronald Press Co., N. Y.

Guberlet, Muriel. 1956. Seaweeds at ebbtide. Univ. of Wash. Press, Seattle.

Hale, Mason E., Jr. 1961. Lichen handbook. Smithsonian Inst., Washington, D. C.

Jackson, Daniel F. 1964. Algae and man. Plenum Press, Inc., N. Y.

Johnson, T. W., and F. K. Sparrow. 1961. Fungi in oceans and estuaries. Hafner Publ. Co., N. Y.

Krieger, Louis C. 1947. The mushroom handbook. Macmillan Co., N. Y.

Lange, Morten, and F. Bayard Hora. 1963. Guide to mushrooms and toadstools. E. P. Dutton and Co., N. Y.

Overholts, Lee O. 1953. The Polyporaceae of the United States, Alaska, and Canada. Univ. of Michigan Press, Ann Arbor.

Prescott, G. W. 1954. How to know the freshwater algae. W. C. Brown Co., Dubuque.

Ramsbottom, John. 1953. Mushrooms and toadstools. Macmillan Co., N. Y.
Smith, Alexander. 1963. The mushroom hunter's field guide. Rev. ed. Univ. of Michigan Press, Ann Arbor.
Smith, Gilbert M. 1950. Freshwater algae of the United States. 2nd ed. McGraw-Hill Book Co., N. Y.
Sparrow, Frederick K., Jr., 1960. Aquatic Phycomycetes. 2nd ed. Univ. of Michigan Press, Ann Arbor.
Taylor, W. R. 1957. Marine algae of the northeastern coast of North America. 2nd ed. Univ. of Michigan Press, Ann Arbor.
Taylor, W. R. 1960. Marine algae of the eastern tropical and subtropical coasts of the Americas. Univ. of Michigan Press, Ann Arbor.
Thomas, William S. 1948. Field book of common mushrooms. G. P. Putnam's Sons, New York.
Tiffany, Lewis H. 1958. Alage. C. C. Thomas, Publ., Springfield, Ill.

MOSSES, LIVERWORTS AND FERNS

Bodenberg, E. T. 1954. Mosses: a new approach to the identification of common species. Burgess Publ. Co., Minneapolis.
Cobb, Boughton. 1956. A field guide to the ferns. Houghton Mifflin Co., Boston.
Conard, Henry S. 1956. How to know the mosses and liverworts. Wm. C. Brown Co., Dubuque, Iowa.
Durand, Herbert. 1949. Fieldbook of common ferns. G. P. Putnam's Sons, New York.
Frye, Theodore C. 1934. Ferns of the Northwest. Binfords and Mort, Portland, Ore.
Frye, Theodore C., and Lois Clark. 1937-1947. Hepaticae of North America. 5 pts. Univ. of Wash. Press, Seattle.
Grout, A. J. 1924. Mosses with a hand lens. Published by the author, Newfane, Vt.
Richards, Paul W. 1950. Mosses. Penguin Books, London.
Smith, Gilbert M. 1955. Cryptogamic botany, Vol. II:Bryophytes and Pteridophytes. McGraw-Hill Book Co., New York.
Watson, E. V. 1964. The structure and life of Bryophytes. Hilary House, New York.
Wherry, E. T. 1948. Guide to eastern ferns. U. of Penn. Press, Philadelphia.
Wiley, Farida. 1948. Ferns of northeastern United States. Amer. Mus. of Nat. Hist., New York.

FLOWERING PLANTS

Abrams, LeRoy. 1923-1960. An illustrated flora of the Pacific States, Vol. I-IV. Stanford Univ. Press, Stanford, Calif.
Anderson, J. P. 1944-1952. Flora of Alaska and adjacent parts of Canada. Iowa State Journal of Sci. Published in numerous parts in Vols. 18-26.
Arnberger, L. P., and J. R. Janish. 1952. Flowers of the Southwest Mountains. Southw. Monuments Assoc., Globe, Ariz.

Craighead, John J., Frank C. Craighead and R. J. Davis. 1963. A field guide to Rocky Mountain wildflowers. Houghton Mifflin Co., Boston.

Dawson, E. Yale. 1963. How to know the cacti. Wm. C. Brown Co., Publ., Dubuque, Iowa.

Enari, Leonid. 1956. Plants of the Pacific Northwest. Binfords and Mort, Publs., Portland, Ore.

Fassett, Norman C. 1957. A manual of aquatic plants. Rev. ed. Univ. of Wisc. Press, Madison.

Fernald, M. 1950. Gray's manual of botany, 8th ed. American Book Co., New York.

Gleason, Henry A. 1952. New Britton and Brown Illustrated Flora of the northeastern United States and adjacent Canada. 3 vols. New York Botanical Garden, New York.

_________________, and Arthur Cronquist. 1963. Manual of vascular plants of northeastern United States and adjacent Canada. D. VanNostrand Co., Inc., Princeton, N. J.

Haskin, Leslie L. 1934. Wildflowers of the Pacific Coast. Binfords and Mort, Portland, Ore.

Hitchcock, C. Leo, Arthur Cronquist, Marion Ownbey and L. W. Thompson 1955-1964. Vascular plants of the Pacific Northwest. 4 pts. Univ. of Washington Press, Seattle.

Jaeger, E. C. 1941. Desert wild Flowers. Stanford Univ. Press, Stanford, Calif.

Matthews, F. S. 1955. Fieldbook of American wildflowers. Rev. ed. G. P. Putnam's Sons, New York.

Moldenke, Harold N. 1949. American wild flowers. D. VanNostrand, Princeton, N. Y.

Muenscher, W. C. 1944. Aquatic plants of the United States. Cornell Univ. Press, Ithaca, New York.

Rydberg, P. A. 1954. Flora of the Rocky Mountains and adjacent plains. (Reprint of 1922 ed.) Hafner Publ. Co., New York.

_______________1932. Flora of the prairies and plains of central North America. N. Y. Botanical Garden, New York.

Sharples, A. W. 1938. Alaska wild flowers. Stanford Univ. Press, Stanford, Calif.

Stewart, Albert, et al.,1963. Aquatic plants of the Pacific Northwest. Oregon State Univ. Press, Corvallis.

Wherry, E. T. 1948. The wild flower guide. Doubleday and Co., New York.

GRASSES, ECONOMIC PLANTS, WEEDS, etc.

Archer, Sellers, and Clarence Bunch. 1953. American grass book. Univ. of Okla. Press, Norman.

Hitchcock, A. S. 1950. Manual of the grasses of the United States. 2nd ed., revised by Agnes Chase. U. S. D. A. Misc. Publ. 200.

Jaques, H. E. 1959. How to know the weeds. Wm. C. Brown Co., Dubuque, Iowa.

Kingsbury, John M. 1964. Poisonous plants of the United States and Canada. 3rd ed. Prentice-Hall Inc., N. Y.

Medsger, Oliver P. 1939. Edible wild plants. Macmillan Co., N. Y.

Moore, Alma. 1960. Grasses. Macmillan Co., New York.
Muenscher, W. C. 1940. Poisonous plants of the United States. Macmillan Co., New York.
________ 1942. Weeds. Macmillan Co., New York.
Pohl, Richard W. 1954. How to know the grasses. Wm. C. Brown Co., Dubuque.

WOODY PLANTS

Baerg, Harry J. 1955. How to know the western trees. Wm. C. Brown Co., Dubuque, Iowa.
Benson, L., and R. A. Darrow. 1952. The trees and shrubs of the southwestern deserts. Univ. of Arizona Press, Tucson, and Univ. of N. Mexico Press, Albuquerque, N. Mex.
Curtis, Carlton C., and S. C. Bausor. 1963. The complete guide to North American trees. Collier Books, New York.
Graves, Arthur H. 1956. An illustrated guide to trees and shrubs. Harper Bros., New York.
Harlow, W. M. 1957. Trees of the eastern United States and Canada. Dover Publ., Inc., New York (reprint edition).
________ 1959. Fruit key and twig key to trees and shrubs. Dover Publ., Inc., New York (reprint edition).
McMinn, Howard E., and Evelyn Maino. 1956. Illustrated manual of Pacific Coast trees. Univ. of Calif. Press, Berkeley, Calif.
Millard, N. D., and W. L. Keene. 1934. Native trees of the intermountain region. U. S. Forest Service, Ogden, Utah.
Petrides, George. 1958. Field guide to the trees and shrubs. Houghton-Mifflin Co., Boston, Mass.
Preston, Richard. 1940. Rocky Mountain trees. Iowa State Univ. Press, Ames, Iowa.
________ 1960. North American trees. Iowa State Univ. Press, Ames, Iowa.
Rosendahl, Carl O. 1955. Trees and shrubs of the upper midwest. Univ. of Minn. Press, Minneapolis, Minn.
Symonds, George W. D., and Stephen Cheminski. 1958. Tree identification book. M. Barrows and Co., New York.
Taylor, R. F. 1929. Pocket guide to Alaska trees. U. S. D. A. Misc. Publ. 55.
VanDersal, W. R. 1938. Native woody plants of the United States. U. S. D. A. Misc. Publ. 303.
Vines, Robert A. 1960. Trees, shrubs and woody vines of the southwest. Univ. of Texas Press, Austin.

INVERTEBRATES

Borror, D. J., and D. M. DeLong. 1964. An introduction to the study of insects. 2nd ed. Holt, Rinehart and Winston, Inc., New York.
Brown, E. S. 1955. Life in fresh water. Oxford Univ. Press, New York.
Buchsbaum, Ralph. 1948. Animals without backbones. Univ. of Chicago Press, Chicago, Ill.

Chu, H. 1949. How to know the immature insects. William C. Brown Co., Dubuque, Iowa.

Dales, R. Phillips. 1963. Annelids. Hillary House, New York.

Eddy, Samuel, and A. C. Hodson. 1961. Taxonomic keys to the common animals of the north central states, exclusive of the parasitic worms, insects and birds. Burgess Publ. Co., Minneapolis Minn.

Edmondson, W. T., H. B. Ward and G. W. Whipple. 1959. Freshwater biology, 2nd ed. John Wiley and Sons, N. Y.

Essig, Edward O. 1958. Insects and mites of western North America. 2nd ed. Macmillan Co., New York.

Green, James. 1963. A biology of Crustacea. Quadrangle Books, Chicago.

Guberlet, Muriel L. 1962. Animals of the seashore. 3rd ed. Binfords and Mort. Publs., Portland, Oregon.

Hyman, Libbie. 1940-1959. The invertebrates, Vols. 1-5. McGraw-Hill Book Co., New York.

Jahn, T. L. 1950. How to know the Protozoa. Wm. C. Brown Co., Dubuque, Iowa

Kaston, B. J., and Elizabeth Kaston. 1953. How to know the spiders. Wm. C. Brown Co., Dubuque, Iowa.

Macan, Thomas T. 1959. Guide to freshwater invertebrate animals. Longmans Green and Co., New York.

Macginitie, George, and Nettie Macginitie. 1949. Natural history of marine animals. McGraw Hill Book Co., New York.

Manwell, Reginald D. 1961. Introduction to protozoology. St. Martin's Press, Inc., New York.

Miner, Roy W. 1950. Field book of seashore life. G. P. Putnam's Sons, New York.

Morris, Percy. 1951. Field guide to the shells of our Atlantic and Gulf coasts. Houghton Mifflin Co., Boston, Mass.

_______________ 1952. Field guide to the shells of the Pacific Coast and Hawaii. Houghton Mifflin Co., Boston, Mass.

Morton, J. E. 1963. Molluscs. Hutchinson Univ. Library, London.

Needham, James G., and Paul R. Needham. 1962. Guide to the study of freshwater biology (reprint edition). Holden-Day Inc., San Francisco, Calif.

Pennak, Robert. 1953. Freshwater invertebrates of the United States Ronald Press, New York.

Pilsbry, Henry A. 1939-1948. Land Mollusca of North America. 4 vols. Monograph 3, Acad. of Nat. Sci. of Philadelphia, Philadelphia, Pa.

Ricketts, E. F., and Jack Calvin. 1952. Between Pacific tides. 3rd ed., rev. by Joel Hedgpeth. Stanford Univ. Press, Stanford, Calif.

Savory, Theodore. 1964. Arachnida. Academic Press, New York. New York.

Schmitt, Waldo L. 1964. Crustacea. Univ. of Michigan Press, Ann Arbor, Mich.

Smith, F. G. W. 1948. Atlantic reef corals. Univ. of Miami Press, Miami, Florida.

Swain, Ralph. 1948. The insect guide. Doubleday and Co., New York.

Tinker, Spencer W. 1964. Pacific Crustacea. Chas. E. Tuttle Co., Rutland, Vermont.

Warmke, Germaine L., and R. T. Abbott. 1962. Caribbean seashells. Livingston Publ. Co., Narberth, Pa.

VERTEBRATES

General references (some include selected invertebrate groups)

Benton, Allen H., and Margaret M. Stewart. 1964. Keys to the vertebrates of the northeastern states, excluding birds. Burgess Publ. Co., Minneapolis, Minn.

Blair, W. F., et al. 1965. Vertebrates of the United States. 2nd ed. McGraw-Hill Book Co., New York.

Collins, Henry H. 1959. Complete field guide to American wildlife. Harper and Row, Publishers, New York.

Eddy, Samuel, and A. C. Hodson. 1961. Taxonomic keys to the common animals of the north central states, exclusive of the parasitic worms, insects and birds. Burgess Publ. Co., Minneapolis, Minn.

Leopold, A. S. and F. Fraser Darling. 1953. Wildlife in Alaska. Ronald Press, New York.

FISHES

Ackermann, William. 1952. Handbook of fishes of the Atlantic seaboard. Greenberg Publ., New York.

American Fisheries Society. 1960. A list of common and scientific names of fishes from the United States and Canada. 2nd ed. American Fisheries Society, McLean, Va.

Bigelow, H. B., et al. 1948-1964. Fishes of the western North Atlantic. Parts 1-4. Memoir No. 1, Bingham Oceanographic Lab., Yale Univ., New Haven, Conn.

Breder, C. M., Jr. 1948. Field book of marine fishes of the Atlantic Coast, G. P. Putnam's Sons, New York.

Eddy, Samuel. 1957. How to know the freshwater fishes. Wm. C. Brown Co., Dubuque, Iowa.

Fisheries Research Board of Canada. 1961. Fishes of the Pacific Coast of Canada. Bulletin No. 68 (2nd ed.) of the Fisheries Research Board of Canada, Ottawa.

Hubbs, Carl, and Karl Lagler. 1960. Fishes of the Great Lakes Region. 2nd ed. Cranbrook Inst. of Sci., Bloomfield Hills, Michigan.

Norman, J. R. 1951. A history of fishes. A. A. Wyn Inc., New York (reprint edition).

Perlmutter, Alfred. 1961. Guide to marine fishes. New York Univ. Press, New York.

Roedel, Phil M. 1953. Common ocean fishes of the California coast. Calif. Dept. of Fish and Game. Fish Bulletin No. 91. Sacramento, Calif.

Schultz, Leonard, and Edith Stern. 1948. The ways of fishes. D. VanNostrand Co., Princeton, N. J.

Scott, W. B. 1954. Freshwater fishes of eastern Canada. Univ. of Toronto Press, Toronto.

Slastenko, E. P. 1958. The freshwater fishes of Canada. Kiev Printers, Toronto, Canada.

Sterba, Gunther. 1963. Freshwater fishes of the world. The Viking Press, New York.
Trautman, Milton B. 1957. The fishes of Ohio. Ohio State Univ. Press, Columbus, Ohio.
Zim, H. S., and H. H. Shoemaker. 1957. Fishes. Golden Press, New York.

AMPHIBIANS AND REPTILES

Barbour, Thomas. 1934. Reptiles and amphibians. Houghton Mifflin Co., Boston, Mass.
Bishop, Sherman. 1943. Handbook of salamanders. Cornell Univ. Press, Ithaca, New York.
Carr, Archie. 1952. Handbook of turtles. Cornell Univ. Press, Ithaca, New York.
Conant, Roger. 1958. A field guide to the reptiles and amphibians. Houghton Mifflin Co., Boston, Mass.
Klauber, Laurence. 1956. Rattlesnakes. 2 vols. Univ. of Calif. Press, Berkeley, Calif.
Logier, E.B.S. 1952. The frogs, toads and salamanders of eastern Canada. Clark Irwin and Co., Ltd., Toronto.
Noble, G. K. 1931. The biology of the Amphibia. McGraw-Hill, N. Y. (Reprinted, 1954, by Dover Publ., N. Y.)
Oliver, J. A. 1955. Natural history of North American amphibians and reptiles. D. VanNostrand Co., Princeton, N. J.
Pope, Clifford. 1939. Turtles of the United States and Canada. Alfred A. Knopf, N. Y.
Pope, Clifford. 1955. The reptile world. Alfred A. Knopf, N. Y.
Schmidt, Karl. 1953. Checklist of the North American amphibians and reptiles. Univ. of Chicago Press, Chicago.
__________, and Dwight D. Davis. 1941. Field book of snakes of the United States and Canada. G. P. Putnam's Sons, N. Y.
Smith, Hobart. 1946. Handbook of lizards of the United States and Canada. Comstock Publ. Co., Ithaca, N. Y.
Stebbins, Robert C. 1954. Amphibians and reptiles of western North America. McGraw-Hill, N. Y.
Wright, A. H., and A. A. Wright. 1949. Handbook of frogs and toads of the United States and Canada. Comstock Publ. Co., Ithaca, N. Y.
____________________ 1957. Handbook of snakes of the United States and Canada. 2 vols. Comstock Publ. Assoc., Ithaca, N. Y.
Zim, Herbert, and Hobart Smith. 1953. Reptiles and amphibians. Golden Press, N. Y.

BIRDS

Allen, A. A. 1961. The book of bird life. 2nd ed. D. VanNostrand Co., Princeton, N. J.
Bent, A. C. 1919 and following. Life histories of North American birds. U. S. Natl. Mus. Bull., Washington.

Booth, Ernest. 1960. Birds of the west. Outdoor Pictures, Escondido, Calif.

__________ 1962. Birds of the east. Outdoor Pictures, Escondido, Calif.

Committee on Nomenclature, American Ornithologists' Union. 1957. Check-list of North American birds. 5th ed. Amer. Ornith. Union, Baltimore, Md.

Gabrielson, Ira N., and F. C. Lincoln. 1959. Birds of Alaska. Stackpole Co., Harrisburg, Pa.

Hausman, Leon A. 1946. Field book of eastern birds. G. P. Putnam's Sons, N. Y.

Hickey, J. J. 1963. A guide to bird watching. Doubleday Anchor Books (reprint edition).

Kortright, F. H. 1942. The ducks, geese and swans of North America. Wildlife Institute, Washington.

Mckenny, Margaret. 1947. Birds in the garden and how to attract them. Univ. of Minn. Press, Minneapolis, Minn.

Munro, George C. 1960. Birds of Hawaii. Chas. E. Tuttle Co., Rutland, Vermont.

Palmer, Ralph S. 1962. Handbook of North American birds. Vol. 1. Yale Univ. Press, New Haven, Conn.

Peterson, Roger T. 1947. A field guide to the birds. Houghton Mifflin Co., Boston, Mass.

__________ 1961. A field guide to western birds. Houghton Mifflin Co., Boston, Mass.

Pettingill, Olin S., Jr. 1951. A guide to bird finding east of the Mississippi. Oxford Univ. Press, N. Y.

__________ 1953. A guide to bird finding west of the Mississippi. Oxford Univ. Press, N. Y.

__________ 1956. A laboratory and field manual of ornithology. 3rd ed. Burgess Publ. Co., Minneapolis, Minn.

Pough, Richard H. 1946. Audubon bird guide: eastern land birds. Doubleday & Co., N. Y.

__________ 1951. Audubon bird guide: water, game and large land birds. Doubleday & Co., N. Y.

__________ 1957. Audubon western bird guide. Doubleday & Co., N. Y.

Saunders, A. A. 1951. A guide to bird songs. Doubleday & Co., N. Y.

__________ 1954. The lives of wild birds. Doubleday & Co., N. Y.

Van Tyne, Josselyn, and Andrew Berger. 1959. Fundamentals of ornithology. John Wiley and Sons, N. Y.

Wallace, George J. 1955. An introduction to ornithology. Macmillan Co., N. Y.

Wolfson, Albert, ed. 1955. Recent advances in avian biology. Univ. of Illinois Press, Urbana, Ill.

MAMMALS

Anthony, Harold. 1928. Field book of North American mammals. G. P. Putnam's Sons, N. Y.

Booth, Ernest S. 1950. How to know the mammals. Wm. C. Brown, Co., Dubuque, Iowa.
Bourliere, Francois. 1954. The natural history of mammals. Alfred A. Knopf Co., New York.
Burt, William. 1957. Mammals of the Great Lakes region. Univ. of Mich. Press, Ann Arbor, Mich.
__________, and Richard Grossenheider. 1952. A field guide to the mammals. Houghton Mifflin Co., Boston, Mass.
Burton, Maurice. 1964. Systematic dictionary of mammals of the world. Thos. Y. Crowell Co., New York.
Cahalane, Victor H. 1947. Mammals of North America. The Macmillan Co., New York.
Cameron, Austin W. 1956. A guide to eastern Canadian mammals. Natl. Mus. of Canada, Ottawa.
Cockrum, E. L. 1962. Introduction to mammalogy. Ronald Press Co., New York.
Davis, David E., and Frank B. Golley. 1963. Principles in mammalogy Reinhold Publ. Corp., New York.
Glass, Bryan. 1951. A key to the skulls of North American mammals. Burgess Publ. Co., Minneapolis, Minn.
Hall, E. Raymond, and Keith Kelson. 1959. The mammals of North America. 2 vols. Ronald Press Co., New York.
Hamilton, W. J., Jr. 1943. The mammals of eastern United States. Cornell Univ. Press, Ithaca, New York.
Miller, G. S., Jr., and Remington Kellogg. 1955. List of North American recent mammals. U. S. Natl. Mus. Bull. 205, Washington, D. C.
Murie, Olaus. 1954. A field guide to animal tracks. Houghton Mifflin Co., Boston, Mass.
Olin, George, and Jerry Cannon. 1959. Mammals of the southwest desert. 2nd ed. Southwestern Monument Assoc., Globe, Ariz.
Palmer, Ralph S. 1954. The mammal guide. Doubleday & Co., New York.
Seton, Ernest Thompson. 1954. Lives of game animals. 8 vols. Doubleday & Co., New York (reprint edition).
Walker, Ernest P. 1964. Mammals of the world. 3 vols. Johns Hopkins Press, Baltimore.
Zim, Herbert, and Donald Hoffmeister. 1955. Mammals. Golden Press, New York.

Audio-visual Aids

Films and filmstrips as well as records offer an opportunity for indoor laboratory work in the winter or during inclement weather. In essence, they may provide an "indoor field trip". The AIBS record on sounds and communications is a worthwhile subject for an indoor laboratory on animal behavior. Use of a laboratory period for these aids is preferable to regular class, since it allows repetition and time for discussion. (It is usually a good idea to show a film twice - once to observe without taking notes, the second time for closer observation and note-taking.)

The following films, filmstrips and records are a selected list, and no attempt was made to be complete. Among the films listed, the authors especially like the Encyclopedia Brittanica series, since they combine good photography and fundamental principles. The McGraw-Hill series is also very good, although the photography is so good it tends to detract from the story being told. The AIBS films were made for high school use, but the level of presentation is generally high enough to be usable at lower collegiate levels as well. They tend to be more filmed lectures than filmed field or laboratory work, however.

Of the film strips listed, the Life series is taken from the Life publication "The World We Live In". They are especially useful on the collegiate level if the instructor makes additional comments with the scenes. The McGraw-Hill film strips were made for use with Woodbury's "Principles of General Ecology" text, on a collegiate level.

There are many bird records on the market, of which only a few are listed. (See General Biological Supply House catalogue for a rather extensive listing.) These records, and the ones on amphibians are useful for learning the songs or calls of birds, frogs and toads. They may be used in conjunction with color slides, or by themselves.

The following symbols are used in connection with the listings:

AIBS — American Institute of Biological Sciences Film Series.
Available from: McGraw-Hill Text-Films
330 West 42nd Street
New York, N. Y. 10036.

C — Cornell University Records
124 Roberts Place
Ithaca, N. Y.

EBF — Encyclopedia Brittanica Films
1150 Wilmette Ave.
Wilmette, Illinois.

FON — Federation of Ontario Naturalists
Edwards Gardens,
Don Mills, Ontario, Canada.

HM — Houghton Mifflin, Co.
Boston, Mass.

L — Life Magazine
Time and Life Building
Rockefeller Center
New York, N. Y.

MH McGraw-Hill Text-Films
330 West 42nd Street
New York, N. Y. 10036

NET NET Film Service
Audio-Visual Center, Indiana University
Bloomington, Indiana.

All films may be purchased from the above-named companies or rented from various education agencies, such as the Audio-visual Center at Indiana Univ. Life film strips may be purchased from the address above for $6.00 apiece, or $5.00 apiece when 4 or more are purchased at one time. McGraw-Hill film strips cost $6.50 apiece or $35.00 for all six.

GENERAL

Films

What is ecology? 11 minutes. EBF
The community. 11 minutes. EBF
The cave community. 13 minutes. EBF
Energy relations. Guest lecturer, Dr. Derry Koob, Wellesley College. 28 minutes. AIBS
Limiting factors. Guest lecturer, Dr. John Vernberg, Duke Univ. 28 minutes. AIBS
Radiation biology. Guest lecturer, Dr. Bert Thomas, Colorado College. 28 minutes. AIBS
Applied ecology. Guest lecturer, Dr. Edward Graham, USDA. 28 minutes. AIBS

Filmstrips

The physical environment. MH
The city as a community. MH

TAXONOMY

Films

The cave community. 13 minutes. EBF

Filmstrips

The enchanted isles. L

TERRESTRIAL COMMUNITIES AND SUCCESSION

Films

Succession- from sand dune to forest. 16 minutes. EBF
The changing forest. 19 minutes. MH
The temperate deciduous forest. 17 minutes. EBF
Life in the woodlot. 17 minutes. MH
The deciduous forest. Guest lecturer, Dr. Lorus Milne. 28 minutes. AIBS
The grasslands. 17 minutes. EBF
The grasslands and the desert. Guest lecturer, Dr. Richard Beidleman, Colorado College. 28 minutes. AIBS
The desert. 22 minutes. EBF
The tropical rain forest. 17 minutes. EBF
The tropical rain forest. Guest lecturer, Dr. Margery Milne. 28 minutes. AIBS
Above the timberline. 16 minutes. MH
Tundra ecology. Guest lecturer, Dr. Richard S. Miller, Univ. of Saskatchewan. 28 minutes. AIBS

Filmstrips

Ecological succession. MH
The forest as a community. MH
The field as a community. MH
The woods of home. L
The desert. L
The rain forest. L
Arctic tundra. L

AQUATIC COMMUNITIES AND SUCCESSION

Films

Limnology. Guest lecturer, Dr. Derry Koob, Wellesley College. 28 minutes. AIBS
World in a marsh. 22 minutes. MH
The spruce bog. 23 minutes. MH
The sea. 26 minutes. EBF
Plankton and the open sea. 19 minutes. EBF
Marine ecology. Guest lecturer, Dr. George Clarke, Harvard Univ. 28 minutes. AIBS
Where land and water meet. (Ecology of the seashore) 30 minutes. NET
On the rocks. (Ecology of the ocean floor) 30 minutes. NET
Life on the coral reef. 30 minutes. NET
Life cycle of the sea. 30 minutes. NET

Filmstrips

The pond as a community. MH
Creatures of the sea. L
Coral reef. L

POPULATION STUDIES

Films

Plant and animal distribution. Guest lecturer, Dr. Derry Koob, Wellesley College. 28 minutes. AIBS
Population ecology. Guest lecturer, Dr. Lawrence Slobodkin, Univ. of Michigan. 28 minutes. AIBS

BEHAVIOR STUDIES

Films

Social insects: the honeybee. 24 minutes. EBF
Behavior. Guest lecturer, Dr. Thomas Eisner, Cornell Univ. 28 minutes. AIBS

Records

Animal sounds and communication. Available only as a demonstration record accompanying the book, "Animal sounds and communication", edited by Lanyon and Tavloga. Available from A. I. B. S., 3900 Wisconsin Ave., N. W. Washington, D. C.
Songs of spring. Songs of 25 birds, repeated frequently enough to be used by the beginner to learn bird songs. FON
American bird songs. Vols. I and II. Vol. I contains songs of 60 birds, Vol. II 51 additional species. C
A field guide to bird songs. A field guide to western bird songs. The first volume named contains 2 records, the second volume 3 records. Each bird song is played briefly after identification, is therefore more suitable for the advanced bird watcher. HM
Voices of the night. Voices of 34 species of frogs and toads. C